Data Science for Infectious Disease Data Analytics

Data Science for Infectious Disease Data Analytics: An Introduction with R provides an overview of modern data science tools and methods that have been developed specifically to analyze infectious disease data. With a quick start guide to epidemiological data visualization and analysis in R, this book spans the gulf between academia and practices providing many lively, instructive data analysis examples using the most up-to-date data, such as the newly discovered coronavirus disease (COVID-19).

The primary emphasis of this book will be the data science procedures in epidemiological studies, including data wrangling, visualization, interpretation, predictive modeling, and inference, which is of immense importance due to increasingly diverse and nonexperimental data across a wide range of fields. The knowledge and skills readers gain from this book are also transferable to other areas, such as public health, business analytics, environmental studies, or spatiotemporal data visualization and analysis in general.

Aimed at readers with an undergraduate knowledge of mathematics and statistics, this book is an ideal introduction to the development and implementation of data science in epidemiology.

Key Features:
- Describes the entire data science procedure of how the infectious disease data are collected, curated, visualized, and fed to predictive models, which facilitates effective communication between data sources, scientists, and decision-makers.
- Describes practical concepts of infectious disease data and provides particular data science perspectives.
- Overview of the unique features and issues of infectious disease data and how they impact epidemic modeling and projection.
- Introduces various classes of models and state-of-the-art learning methods to analyze infectious diseases data with valuable insights on how different models and methods could be connected.

Dr. Lily Wang is a tenured professor of statistics at George Mason University. She earned her PhD in statistics from Michigan State University in 2007. Before joining Mason in 2021, she was on the faculty of Iowa State University (2014-2021) and the University of Georgia (2007-2014). Her primary research areas include non/semi-parametric modeling and inference, statistical learning of data objects with complex features, methodologies for functional data, spatiotemporal data, imaging, and general issues related to data science and big data analytics. Dr. Wang is a fellow of both the Institute of Mathematical Statistics and the American Statistical Association and an Elected Member of the International Statistical Institute. She is currently serving on the editorial board of *Journal of the Royal Statistical Society, Series B, Journal of Nonparametric Statistics*, and *Statistical Analysis and Data Mining*.

CHAPMAN & HALL/CRC DATA SCIENCE SERIES

Reflecting the interdisciplinary nature of the field, this book series brings together researchers, practitioners, and instructors from statistics, computer science, machine learning, and analytics. The series will publish cutting-edge research, industry applications, and textbooks in data science.

The inclusion of concrete examples, applications, and methods is highly encouraged. The scope of the series includes titles in the areas of machine learning, pattern recognition, predictive analytics, business analytics, Big Data, visualization, programming, software, learning analytics, data wrangling, interactive graphics, and reproducible research.

Published Titles

Explanatory Model Analysis
Explore, Explain, and, Examine Predictive Models
Przemyslaw Biecek, Tomasz Burzykowski

An Introduction to IoT Analytics
Harry G. Perros

Data Analytics
A Small Data Approach
Shuai Huang and Houtao Deng

Public Policy Analytics
Code and Context for Data Science in Government
Ken Steif

Supervised Machine Learning for Text Analysis in R
Emil Hvitfeldt and Julia Silge

Massive Graph Analytics
Edited by David Bader

Data Science
An Introduction
Tiffany-Anne Timbers, Trevor Campbell and Melissa Lee

Tree-Based Methods
A Practical Introduction with Applications in R
Brandon M. Greenwell

Urban Informatics
Using Big Data to Understand and Serve Communities
Daniel T. O'Brien

Data Science for Infectious Disease Data Analytics
An Introduction with R
Lily Wang

For more information about this series, please visit: https://www.routledge.com/
Chapman--HallCRC-Data-Science-Series/book-series/CHDSS

Data Science for Infectious Disease Data Analytics
An Introduction with R

Lily Wang

CRC Press
Taylor & Francis Group
Boca Raton London New York

CRC Press is an imprint of the
Taylor & Francis Group, an **informa** business

A CHAPMAN & HALL BOOK

First edition published 2023
by CRC Press
6000 Broken Sound Parkway NW, Suite 300, Boca Raton, FL 33487-2742

and by CRC Press
4 Park Square, Milton Park, Abingdon, Oxon, OX14 4RN

CRC Press is an imprint of Taylor & Francis Group, LLC

© 2023 Lily Wang

Library of Congress Cataloging-in-Publication Data

Names: Wang, Lily (Professor of Statistics), author.
Title: Data science for infectious disease data analytics : an introduction with R / Lily Wang.
Description: First edition. | Boca Raton : CRC Press, 2023. | Series: Chapman & Hall/CRC data science series | Includes bibliographical references and index.
Identifiers: LCCN 2022011249 (print) | LCCN 2022011250 (ebook) | ISBN 9781032187426 (hbk) | ISBN 9781032188058 (pbk) | ISBN 9781003256328 (ebk)
Subjects: LCSH: Epidemiology--Statistical methods. | Epidemiology--Data processing. | R (Computer program language)
Classification: LCC RA652.2.M3 W36 2023 (print) | LCC RA652.2.M3 (ebook) | DDC 614.4--dc23/eng/20220322
LC record available at https://lccn.loc.gov/2022011249
LC ebook record available at https://lccn.loc.gov/2022011250

ISBN: 978-1-032-18742-6 (hbk)
ISBN: 978-1-032-18805-8 (pbk)
ISBN: 978-1-003-25632-8 (ebk)

DOI: 10.1201/9781003256328

Typeset in Latin Modern
by KnowledgeWorks Global Ltd.

Publisher's note: This book has been prepared from camera-ready copy provided by the authors.

To all people, teams, and communities who have devoted their efforts to confronting COVID-19.

Contents

Preface

The book will be an excellent reference for four types of audiences: (1) data scientists with interest in statistical analysis of epidemiological data; (2) undergraduate and graduate students studying biostatistics or epidemiology; (3) epidemiologists learning R to work on data science issues; and (4) practitioners learning epidemic modeling in general and developing data science tools for epidemiological data.

In most chapters, we assume that readers are familiar with introductory statistics. A couple of chapters also require knowledge of calculus, for example, epidemic modeling and neural networks.

Built from the ground up for statistical analysis, R has become one of the favorite programming languages for data scientists. There are many reasons why the R programming language has been so popular in data science. Several important reasons include the open-source data operation packages and utilities, various statistical and graphical capabilities, a wide range of database support, and interactive web-based dashboards. We use the R programming language throughout the book, and we intend for students to learn how to use R for data visualization, modeling, and forecasting, especially in epidemiology. See Appendix A for instructions on installing and using R.

Short Description: This book provides a quick start guide to infectious disease data visualization and analysis in R. Readers will learn how to handle infectious disease data from data science perspectives.

Long Description: This book will introduce readers to modern statistical models and state-of-the-art learning methods to analyze infectious disease data. Many key approaches, examples, and case studies will be presented. The primary emphasis will be the hands-on data analyses, including data visualization, interpretation, epidemic modeling, predicting, and inference. This book also provides practical concepts and R programming skills to perform infectious disease data analysis and visualization.

Specifically, readers will learn how to:

- develop the ability to use R language to understand basic features of the data, fundamental data processing skills, and primary data visualization skills;

- apply appropriate descriptive and inferential statistical techniques to infectious disease data and interpret results of statistical analyses in the context of public health research and evaluation;
- develop and present of results of statistical analysis of epidemiological data;
- develop skills in model building, investigating model assumptions, and interpreting modeling and forecasting results with particular applications to epidemiologic and especially the infectious disease data.

Key Features of the Book

- This book describes the entire data science procedure of how the infectious disease data are collected, curated, visualized, and fed to predictive models, which facilitates effective communication between data sources, scientists, and decision-makers.

- The book describes practical concepts of infectious disease data and provides particular data science perspectives.

- This book gives an overview of the unique features and issues of infectious disease data and how they impact epidemic modeling and projection.

- The book introduces various classes of models and state-of-the-art learning methods to analyze infectious disease data. In addition, it provides valuable insights on how different models could be connected.

- This book spans the gulf between academia and practices, providing many hands-on data analysis examples using the most up-to-date data, such as the newly discovered COVID-19.

- The book focuses on data wrangling and visualizations, which are ubiquitous due to increasingly diverse and unstructured data across a wide range of fields. The skills users gain from this book are also transferable to other areas, such as public health studies, business, time series, or space-time data analysis in general.

Datasets Used in Examples and Exercises

In this book, we illustrate statistical learning methods for infectious disease data using applications from COVID-19 data. There are end-of-chapter exercises for most of the chapters, many of which involve the use of R packages to study the datasets provided. The IDDA package available on the book website contains many datasets required to implement all the R code from the examples and work on exercises associated with this book.

Book Webpage

A webpage (`https://first-data-lab.github.io/IDDA_book/`) is also created for this book for demonstrating the colored version of the figures, animations, shiny apps, and other interactive visualizations of the examples presented in this book.

Acknowledgments

I genuinely appreciate all the contributors to the open-source projects of R and RStudio. I couldn't have made this book without their contributions and the packages that they developed. I am immensely grateful to the members of the FIRST Data Science Lab including Dr. Myungjin Kim, Miss Zhiling Gu, Dr. Guannan Wang[1], Dr. Xinyi Li[2], Dr. Shan Yu[3], and Dr. Yueying Wang for their help with the book. This book wouldn't be possible without their generous assistance. In addition, we would like to express my appreciation to the editor, Dr. David Grubbs, and the anonymous reviewers for their valuable feedback and suggestions. I wish to thank Iowa State University and George Mason University for their continuing support during the writing of this book. I would also like to thank Dr. Yihui Xie for his "rmarkdown" package[4], which simplifies the writing of this book by having all content written in R Markdown.

Last but not least, I want to thank my family. Balancing work, family, and life amidst the COVID-19 pandemic is very challenging. Writing and finishing this book in the pandemic added a significant time commitment to my usual heavy workload. It also requires much concentration. There were many days when I had to work late or even overnight to write this book. I want to thank my entire family for their support, and especially I would like to thank my three

---

[1]https://people.wm.edu/~gwang01/index.html

[2]https://mthsc.clemson.edu/directory/view_person.py?person_id=734

[3]https://statistics.as.virginia.edu/faculty-staff/profile/sy5jx

[4]https://cran.r-project.org/web/packages/rmarkdown

lovely kids, Annie, Andrew, and Angie, for their scheduled and unscheduled interruptions so I could take breaks from the computer screen.

About the Author

Dr. Lily Wang is a tenured professor of statistics at George Mason University. She earned her PhD in statistics from Michigan State University in 2007. Before joining Mason in 2021, she was on the faculty of Iowa State University (2014–2021) and the University of Georgia (2007–2014). Her primary research areas include: non-/semi-parametric modeling and inference, statistical learning of data objects with complex features, methodologies for functional data, spatiotemporal data, imaging, survey sampling, and data reduction methods. Working at the interface of statistics, mathematics, and computer science, she is also interested in general issues related to data science and big data analytics. The methods she developed have a wide application in engineering, neuroimaging, epidemiology, environmental studies, economics, and biomedical science. Dr. Wang is a fellow of both the Institute of Mathematical Statistics and the American Statistical Association and an Elected Member of the International Statistical Institute.

1

Introduction

1.1 Aims and Scope of This Book

An epidemiologic investigation is arguably where the presence of data is most imperative and critical, as this is when an issue of the health of the public is at the highest stake and demands rapid intervention. Epidemiological data is imperative to enforce measures that ensure that the public's health and safety are preserved. However, in the case that one would conduct a field investigation, relevant existing data must exist. If not, then the bottom line would be to compile a new set of data that focuses on the target of an investigation.

In light of the advancements of this current society, it is not about the data or the lack thereof which creates hardships, rather the act of determining which data is the most significant and better suited to show the most useful and valid results. This pushes one to determine how to collect data from a variety of sources that may be difficult to analyze for a handful of field epidemiological investigations. Infectious disease learning requires lots of emphasis on data manipulation, visualization and modeling. This book aims to provide an overview of modern data science methods and tools that have been explicitly developed to analyze, interpret and present infectious disease data. Readers are assumed to have a background in introductory-level statistics and multivariate calculus (for some chapters), but no specialist knowledge of infectious diseases. Due to the size of this topic, we do not claim to provide comprehensive coverage of all existing methods. However, we will describe many of the critical approaches, and throughout, there will be many examples and case studies. This book serves the complementary purposes of (i) introducing graduate students and others to the field of infectious disease data analysis, (ii) acting as a reference for researchers in this field, and (iii) helping practicing data scientists and infectious disease epidemiologists develop the ability to use the R programming language to understand basic data features, perform data processing and visualization, and build epidemic modeling and forecasting skills.

1.2 The Structure of This Book

1.2.1 Infectious Disease Data

The material in this book is concerned with the statistical analysis of quantitative data obtained by observing the spread of **infectious diseases**. According to Porta (2014):

> Infectious disease (or communicable disease) is defined as an illness caused by a specific infectious agent or its toxic product that results from transmission of that agent or its products from an infected person, animal, or reservoir to a susceptible host, either directly or indirectly through an intermediate plant or animal host, vector or inanimate environment. — Porta (2014)

Public health officials analyze the infectious diseases that are either ongoing or emerging by utilizing surveillance systems. Surveillance systems compile data to help with that. Stabilizing a health issue will be significantly more difficult without a thorough understanding of whatever the health problem (etiology, distribution, mechanism of infection, etc.) may be.

Coronavirus disease (COVID-19) is an infectious disease caused by a newly discovered coronavirus. It is well known that data is critical to understand the impact of infectious disease and inform the appropriate response, planning, and allocation of resources. The COVID-19 pandemic raises essential questions about opening (the availability and accessibility), sharing and using data, as well as highlights the challenges associated with data use.

1.2.2 Basic Characteristics of the Infection Process

Figure 1.1 illustrates the chain of infection presented in Dicker and Gathany (1992).

The **reservoir of infection**, also called the **primary source of infection**, is a locality (person, animal, arthropod, plant, soil, or substance) where an infectious agent is able to thrive and multiply, as well as where the agent is capable of being transmitted to another susceptible host. See Barreto et al. (2006).

The conditions required for a new host to become infected require a circumstance where the susceptible host is able to be exposed to the infectious agent. In simple terms, contact between the agent and the host is necessary for an infection to occur. The infectious disease is transmitted to a susceptible host when the individual takes in a sufficient quantity of causative organisms. When a susceptible host takes in an amount of infectious material sufficient to induce infection, we say the individual has made an **infectious contact**. After the infectious contact has occurred, most individuals go through what is called a **latent period**. A latent period is a period of time where the infection exclusively evolves internally, lacking any sort of emission which may be caused by the infectious material. The **latent period** is the time from infection to the onset of the ability to infect. It ends when the infected individual becomes infectious, and for the duration of the infectious period, we refer to the infected individual as an **infective** who can transmit the infection to susceptible hosts.

The infectious period ends when an infected individual is no longer considered infectious. Therefore, they may either become susceptible again or a removal. A **removal** describes an individual who is either immune or deceased and will no longer be able to spread the disease. Each infected individual is also referred to as a **case**.

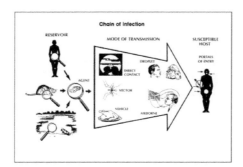

FIGURE 1.1: Chain of infection. Source: Dicker and Gathany (1992). *Principles of Epidemiology*. Public Health Service, Centre for Disease Control and Prevention (CDC), Atlanta, second edition.

1.2.3 Data Visualization

Data visualization or information visualization has always played a crucial role in scientific analysis. In many infectious disease studies, data visualization could be a good starting point for users to understand how far the disease will spread and to illustrate our findings and statistical insights. Besides, the ability to visualize, track, and predict the spread of the disease can help raise awareness, understand the impact of the disease, and ultimately assist in prevention efforts.

Data visualization has become a vital aspect of the COVID-19 pandemic. Social media feeds are overwhelmed with infection heat maps and charts depicting transmission patterns. We have all seen models projecting the spread of the novel coronavirus. Data scientists have met great difficulty with the COVID-19 pandemic due to the nature of the virus, including the vast and rapid spread, as well as the substantial impact on the economy. A lot of work has been done on visualizing COVID-19 data since the outbreak of the pandemic. The daily counts of cases and deaths of COVID-19 are crucial for understanding how this pandemic is spreading. Thanks to the contribution of the data science communities across the world, multiple sources provide COVID-19 data with different precision and focus. To clean the data, we first fetch data from various sources and compile them into the same format for further comparison and integration. Appendix B describes the data used in the examples, case studies, and lab exercises in the book.

Figure 1.2 summarizes the data science procedure to learn the infectious disease data, facilitate effective communication between data sources, scientists and decision-makers, and provide important guidelines for the public. Using this procedure, Kim et al. (2021) explained how the COVID-19 data were collected, curated, visualized, and fed to predictive models in their practice.

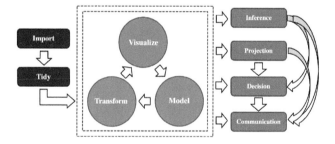

FIGURE 1.2: The stages of a data science workflow. Source: Kim, et al. (2021). Methods, challenges, and practical issues of COVID-19 projection: A data science perspective. *Journal of Data Science*, 19(2).

R gifts users the ability to scale, automate, document and track tasks, as well as reproduce their output in a reliable manner. The first few chapters investigate existing R visualization techniques used to manipulate and represent infectious disease data. Chapter 2 provides an introduction to data wrangling and how to use R packages "dplyr" and "tidyr" to manipulate the data in a helpful form for visualization and modeling. This chapter is for someone who is already somewhat familiar with R but would like to know more about using it for fundamental data analysis and manipulation.

Graphs can be presented using a variety of media: print, projection, dashboard, etc. The visualizations can be distributed into two groups: visualization with zero or less interactivity, and complex interactive techniques and tools. See

Figure 1.3 for different types of visualization. Before constructing any display of epidemiologic data, it is important to first determine the point to be conveyed and which media we want to use for communications.

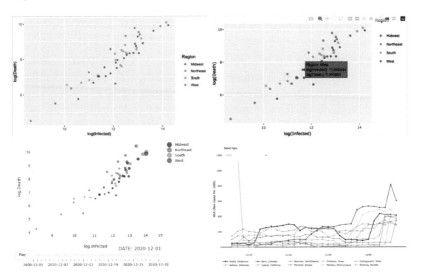

FIGURE 1.3: Types of visualization plots. See our book webpage for an interactive version of these plots.

Chapter 3 introduces static visualization, which uses basic graphs such as bar and line graphs for representing attributes of the COVID-19 dataset. We use a collection of graphs for comparing cumulative or daily new cases and deaths between states/counties in the US.

Chapter 4 provides insight and practical skills for creating interactive and dynamic web graphics for data analysis using R. These kinds of visualizations allow user interaction like hovering the mouse over bars and points in the charts. It makes heavy use of the "plotly" package for rendering graphics, but readers will also learn about other R packages that augment a data science workflow, such as the "tidyverse." Along the way, readers will gain insight into best practices for visualization of infectious disease data, statistical graphics and graphical perception. Chapter 5 focuses on linking "plotly" graphs with "shiny," an open-source R package that provides an elegant and powerful web framework for building web applications. Chapter 6 is an in-depth look at visualizing data in a spatial setting and presenting findings through some geospatial visualization.

1.2.4   Epidemic Modeling and Forecasting

The concepts and techniques discussed in Chapters 2–5 have dealt with describing, visualizing, and exploring the data. The use of scientific models for understanding the dynamics of infectious diseases has a very rich history in epidemiology.

Starting in December 2019 in China, the outbreak of COVID-19 spread globally within weeks. COVID-19 must be fought in a way where scientists and researchers recognize how critical it is to comprehend how far the virus will spread and how many lives it may take. Scientific modeling is an essential tool to answer these questions and ultimately assist disease prevention, policy-making and resource allocation.

Chapter 7 presents a few classic epidemic modeling approaches and takes readers through steps required for fundamental infectious disease data analysis and presentation of data typically encountered in epidemiology using the COVID-19 dataset. Chapter 8 introduces the compartment models. Chapter 10 provides the analytical techniques of regression and discrimination as a means of quantifying the effect of a set of explanatory variables on a particular outcome.

One of the most important types of data one encounters in infectious disease surveillance are epidemiological time series. Chapter 9 takes readers through time series modeling and forecasting. Chapter 11 introduces some neural network models for forecasting. Chapter 12 describes the hybrid models using multiple forecasting algorithms to improve the predictive performance.

Appendix A provides a general introduction to R and some basic programming skills that are generally unrelated to the use of R as statistical software. Appendix B describes the datasets considered in the main chapters of the book. Appendix C describes how to handle dates and times in R.

2

Data Wrangling

An essential part of data science is data wrangling, a process designed to clean and unify messy and complex data, and transform raw data into more readily used formats. However, since the infectious disease data involve unique features not shared by traditional datasets, the importance of managing and leveraging information with data is more emphasized in epidemiology analysis. Practical skills for data wrangling are discussed based on two examples of epidemic data: the COVID-19 infected cases and the CDC outpatient influenza-like illness data. Specifically, this chapter focuses on examining how to transform data efficiently and how to extract and summarize insights from it, which is a vital step for the data science procedure such as visualizing and modeling.

2.1   An Introduction to R Packages "dplyr" and "tidyr"

The package "dplyr" is an R package created to make tabular data wrangling less difficult by using a limited set of functions that can be used together to extract and summarize insights from the data. It pairs well with "tidyr," a tool that allows us to swiftly convert between two data formats (long vs. wide) for graphing and analysis. Various R functions use it to solve the challenge of reshaping data for visualization. We may need datasets with one row per measurement, a data frame with each measurement type having its own column, and rows being more aggregated groupings, or we may need to choose relevant variables, filter out key observations, generate new variables, and obtain summary statistics. As illustrated in Figure 2.1, it is necessary that we work back and forth between those formats. R packages "tidyr" and "dplyr" give us the right tools for this as well as for more advanced and intricate data wrangling.

The packages "dplyr" and "tidyr" are built to work directly with data frames. This chapter will introduce how to use these packages to perform data manipulation and transform the data into the appropriate form.

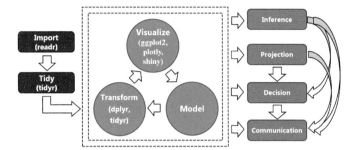

FIGURE 2.1: A typical data science process.

2.2 Learning R Package "dplyr"

2.2.1 Tibbles

Throughout this book, we will mostly be using "tibbles" instead of R's "data.frame". Tibbles are data frames but modernized to keep what time has proven to be effective and throw out what is not. We will utilize the "tibble" package to create opinionated data frames that make working in "tidyverse" a bit simpler.

```
# The easiest way to get tibble is to install the whole tidyverse:
install.packages("tidyverse")

# Alternatively, install just tibble:
install.packages("tibble")
```

For the most part, we will use the terms "tibble" and "data frame" interchangeably. When we choose to point out R's data frame, we will call it a "data.frame". See Wickham and Grolemund (2016) for more details about how to create and use "tibbles."

2.2.2 Importing Data

Recall R offers many ways to import data:

- `read_csv()` reads comma-separated files, and `read_csv2()` reads semicolon-separated files;

- read_tsv() reads tab-separated files, and read_delim() reads files with any delimiter;

- read_fwf() reads fixed-width files. We can use fwf_widths() or fwf_positions() to specify fields by their widths or position, respectively;

- read_table() reads a common variation of fixed-width files where columns are separated by white space;

- load() loads an .RData file and imports all of the objects contained in the .RData file into the current workspace.

Example 2.1. (COVID-19 US County-level Infected Count Data).
We work on a COVID-19 US county-level infected count data (I.county), which can be obtained from the GitHub R package IDDA. It is a data frame which includes ID (county-level Federal Information Processing System code), County (name of county), State (name of state), XYYYY.MM.DD (the number of the cumulative infected cases in a county related to the date of YYYY.MM.DD) for 3,104 counties in the US. For example, the variable X2020.01.22 is the number of the cumulative infected cases in a county on January 22, 2020. See Appendix B for a more detailed description of the data and its source.

```
# Load the packages
library(tidyverse); library(dplyr)
```

```
## -- Attaching packages -------------- tidyverse 1.3.1 --
```

```
## v ggplot2 3.3.5      v purrr   0.3.4
## v tibble  3.1.3      v dplyr   1.0.7
## v tidyr   1.1.3      v stringr 1.4.0
## v readr   2.0.1      v forcats 0.5.1
```

```
## -- Conflicts ---------------- tidyverse_conflicts() --
## x dplyr::filter() masks stats::filter()
## x dplyr::lag()    masks stats::lag()
```

```
# Install the IDDA package from github
library(devtools)
devtools::install_github('FIRST-Data-Lab/IDDA')
```

```
# Load objects in I.county into my workspace
library(IDDA)
data(I.county)
# Make I.county a tibble with as_tibble()
I.county <- as_tibble(I.county)
```

```
# Preview the data
View(I.county)
```

Example 2.2. (The CDC ILI Data). The CDC FluView Portal provides national, regional, and state-level outpatient influenza-like illness data, which is available using the ilinet() function from the "cdcfluview" R package. We will also work on this data. The total number of patients visiting hospitals (total_patients), patients with influenza-like illness (ILI) (ilitotal), and patients with the "ILI" by age group (age_0_4 to age_65) are recorded for each week (week_start). See Appendix B for a more detailed description of the data and its source.

```
# Install cdcfluview package to get ILI data:
install.packages("cdcfluview")
```

```
library(cdcfluview)
Ili.usa <- ilinet(region = "national", years = NULL)
# Make Ili.usa a tibble with as_tibble()
Ili.usa <- as_tibble(Ili.usa)
```

```
# Information dense summary of tbl data
dplyr::glimpse(Ili.usa)
```

```
## Rows: 1,291
## Columns: 16
## $ region_type    <chr> "National", "National", "Nat~
## $ region         <chr> "National", "National", "Nat~
## $ year           <int> 1997, 1997, 1997, 1997, 1997~
## $ week           <int> 40, 41, 42, 43, 44, 45, 46, ~
## $ weighted_ili   <dbl> 1.101, 1.200, 1.379, 1.199, ~
```

```
## $ unweighted_ili    <dbl> 1.217, 1.281, 1.239, 1.145, ~
## $ age_0_4           <dbl> 179, 199, 228, 188, 217, 178~
## $ age_25_49         <dbl> NA, NA, NA, NA, NA, NA, NA, ~
## $ age_25_64         <dbl> 157, 151, 153, 193, 162, 148~
## $ age_5_24          <dbl> 205, 242, 266, 236, 280, 281~
## $ age_50_64         <dbl> NA, NA, NA, NA, NA, NA, NA, ~
## $ age_65            <dbl> 29, 23, 34, 36, 41, 48, 70, ~
## $ ilitotal          <dbl> 570, 615, 681, 653, 700, 655~
## $ num_of_providers  <dbl> 192, 191, 219, 213, 213, 195~
## $ total_patients    <dbl> 46842, 48023, 54961, 57044, ~
## $ week_start        <date> 1997-09-28, 1997-10-05, 199~
```

2.2.3   Common "dplyr" Functions

Next, we will review some of the most frequently used "dplyr" functions:

- select(): subsets columns;
- filter(): subsets rows according to some criteria;
- mutate(): creates new columns by combining data from existing ones;
- summarize(): produce summary statistics on grouped data with group_by();
- slice_max() and slice_min(): select rows with highest or lowest values of a variable;
- arrange(): sort results;
- join() family: combine datasets.

A typical code structure of "dplyr" is:

```
data.new <- data.original %>%
  select rows or columns to manipulate %>%
  arrange or group the data %>%
  summarize the data
```

The first argument is a data frame, and the subsequent arguments separated by %>% describe the data manipulation and/or summary, and the result is a new data frame. We will explain it in more detail in the following sections.

2.3   Selecting Columns and Filtering Rows

2.3.1   Subsetting Variables

We can use `select()` to choose columns from a data frame. The first argument to this function is the data frame (`I.county`), and the subsequent arguments are the columns to keep, separated by commas. Alternatively, we may choose adjacent columns by using a ":" to select a range of columns, read as in "select columns from ____ to ____."

```
# Load the tidyverse
dplyr::select(I.county, ID, County, State)
# Select a series of connected columns
dplyr::select(I.county, ID, County, State, X2020.12.31:X2020.12.22)
```

2.3.2   Subsetting Observations

The `filter()` function may be used to select rows based on certain criteria. The arguments after the data frame are the condition(s) we want for our final data frame to adhere to (for example, `State` name is "Florida").

```
# All Florida counties
dplyr::filter(I.county, State == "Florida")
```

Using commas between conditions, we can chain a sequence of criteria together. Here is an example of the `filter()` function with multiple conditions:

```
# All Florida counties with cumulative infection count > 10000
dplyr::filter(I.county, State == "Florida", X2020.12.31 > 10000)
```

To use filtering effectively, it is better to know some of the comparison and logical operators. Table 2.1 shows some commonly used R logic comparisons:

TABLE 2.1: Some commonly used logic comparisons.

Logics	Descriptions
<	Less than
<=	Less than or equal to
>	Greater than
>=	Greater than or equal to
==	Equal to
!=	Not equal to
%in%	Group membership
is.na	Is NA
!is.na	Is not NA
x & y	X AND y
x | y	x OR y
!x	Not x
any(...)	Are some values true in ...?
all(...)	Are all values true in ...?

2.3.3 Pipes

What if we wanted to select and filter at the same time? There are at least three options available, and one can use intermediate steps, nested functions, or pipes.

With intermediate steps, one can establish a temporary data frame and use it as an input for the following function; for example,

```
# All Florida counties from 2020.12.22 to 2020.12.31
# Method 1
FL.I.county <- dplyr::filter(I.county, State == "Florida")
FL.I.county.DEC <- dplyr::select(FL.I.county,
                        X2020.12.31:X2020.12.22)
```

This is readable, but can easily fill up the workspace with objects that need to be named individually. With multiple steps, that can be difficult to keep track of.

We can also nest functions (i.e., one function inside of another), like this:

```
# All Florida counties from 2020.12.22 to 2020.12.31
# Method 2
```

```
FL.I.county.DEC <-
  dplyr::select(dplyr::filter(I.county, State == "Florida"),
                ID, County, State, X2020.12.31:X2020.12.22)
```

This is useful but can be tricky to read if too many functions are nested, as R evaluates the expression from the inside out (in this case, filtering, then selecting).

The last option, **pipes**, were recently added to R. Pipes allow us to transfer the output of one function directly to the next, which is handy when working with the same dataset numerous times. Pipes in R are represented by the `%>%` symbol and are available through the "magrittr" package, which is installed automatically with "dplyr."

```
# All Florida counties from 2020.12.22 to 2020.12.31
# Method 3
I.county %>%
  dplyr::filter(State == "Florida") %>%
  dplyr::select(ID, County, X2020.12.31:X2020.12.22)
```

In the code above, we use the pipe to pass the I.county data first through `filter()` to keep rows for the state of Florida, then the `select()` function is used to keep only the counts from December 22 to December 31, 2020. Using `%>%`, we can send the object on its left and pass it as the first argument to the function on its right; therefore, we don't need to explicitly include the `data.frame` as an argument to the `filter()` and `select()` functions anymore.

It may be more helpful to read the pipe like the word "then." For example, as shown above, we take the data frame I.county and filter for rows with `State == "Florida"`, then we select columns from `X2020.12.31` to `X2020.12.22`. Although the "dplyr" functions are basic, they may be combined into linear workflows using the pipe to do more complicated data wrangling processes.

If we make a new object with the smaller version of the data, we could give it a new name:

```
# Assign a name to all Florida counties
# from 2020.12.22 to 2020.12.31
FL.I.county.DEC <- I.county %>%
  dplyr::filter(State == "Florida") %>%
  dplyr::select(ID, County, X2020.12.31:X2020.12.22)
```

The function `print()` can be used to show tibble results with the options n and `max_extra_cols` to control how many rows and columns are printed.

```
# head(FL.I.county.DEC)
FL.I.county.DEC %>%
  print(n = 2, max_extra_cols = 5)
```

```
## # A tibble: 67 x 12
##       ID County  X2020.12.31 X2020.12.30 X2020.12.29
##    <int> <chr>         <int>       <int>       <int>
## 1 12001 Alachua       15866       15674       15574
## 2 12003 Baker          2470        2453        2432
## # ... with 65 more rows, and 7 more variables:
## #   X2020.12.28 <int>, X2020.12.27 <int>,
## #   X2020.12.26 <int>, X2020.12.25 <int>,
## #   X2020.12.24 <int>, ...
```

2.3.4 Selecting Rows with Highest or Lowest Values of a Variable

Sometimes we would like to select rows with the highest or lowest values of a variable. The function `slice_max()` (`slice_min()`) can be used to select top (or bottom) n rows (by value).

This is a handy wrapper that uses `filter()` and `min_rank()` to choose the top or bottom entries in each group, ordered by `order_by`.

```
slice_max(x, order_by, n)
slice_min(x, order_by, n)
```

- x: a `tbl()` to filter;
- n: number of rows to return;

- `order_by`: the variable to use for ordering.

Let us find the top ten counties with the largest cumulative infected count on December 31, 2020.

```
# Top 10 counties with the cum. infected count
I.county.top10 <- I.county %>%
  slice_max(order_by = X2020.12.31, n = 10)
I.county.top10$County
```

```
##  [1] "LosAngeles"    "NewYork"       "Cook"
##  [4] "Maricopa"      "Miami-Dade"    "Harris"
##  [7] "SanBernardino" "Dallas"        "Riverside"
## [10] "Clark"
```

Let us find the bottom ten counties with the smallest cumulative infected count on December 31, 2020.

```
# Bottom 10 counties with the cum. infected count
I.county.bottom10 <- I.county %>%
  slice_min(order_by = X2020.12.31, n = 10)
I.county.bottom10$County
```

```
##  [1] "OglalaLakota" "Loving"       "King"
##  [4] "Harding"      "Petroleum"    "Hinsdale"
##  [7] "Blaine"       "Wheeler"      "Arthur"
## [10] "Borden"       "Daggett"
```

Let us find the county with the largest cumulative infected count on December 31, 2020 for each state.

```
# County with the largest cum. infected count for each state
I.county.top1 <- I.county %>%
  group_by(State) %>%
  slice_max(order_by = X2020.12.31, n = 1) %>%
  dplyr::select(State, County)
```

2.3.5   Additional Features

2.3.5.1   Selection Helpers

First, we introduce some of the selection helpers, which are used in `select()`, as follows:

- `starts_with()`: starts with a prefix;
- `ends_with()`: ends with a suffix;
- `contains()`: contains a literal string;
- `one_of()`: select columns whose names are in a group of names;
- `matches()`: matches a regular expression;
- `:`: select all columns between two variables;
- `-`: select all columns except the variable;
- `num_range()`: matches a numerical range like x01, x02, x03;
- `last_col()`: select the last variable.

Here, we start with the ILI example to see how to use the selection helpers.

```
# Select the column name which starts with "age_"
Ili.usa %>% select(starts_with("age_"))
```

```
## # A tibble: 1,291 x 6
##    age_0_4 age_25_49 age_25_64 age_5_24 age_50_64 age_65
##      <dbl>     <dbl>     <dbl>    <dbl>     <dbl>  <dbl>
## 1      179        NA       157      205        NA     29
## 2      199        NA       151      242        NA     23
## # ... with 1,289 more rows
```

```
# Select the column name which ends with "_ili"
Ili.usa %>% select(ends_with("_ili"))
```

```
## # A tibble: 1,291 x 2
##    weighted_ili unweighted_ili
##           <dbl>          <dbl>
## 1          1.10           1.22
## 2          1.20           1.28
## # ... with 1,289 more rows
```

```
# Select the column name which contains "_"
Ili.usa %>% select(contains("_"))
```

```
## # A tibble: 1,291 x 12
##    region_type weighted_ili unweighted_ili age_0_4
```

```
##    <chr>                  <dbl>          <dbl>    <dbl>
## 1 National                1.10           1.22      179
## 2 National                1.20           1.28      199
## # ... with 1,289 more rows, and 8 more variables:
## #    age_25_49 <dbl>, age_25_64 <dbl>, age_5_24 <dbl>,
## #    age_50_64 <dbl>, age_65 <dbl>,
## #    num_of_providers <dbl>, total_patients <dbl>,
## #    week_start <date>
```

```
# Keep "total_patients" without an error
# though there is no "garbage_Variable".
# This is useful when we don't know which columns will be present.
Ili.usa %>% select(one_of(c("total_patients","garbage_variable")))
```

```
## # A tibble: 1,291 x 1
##    total_patients
##             <dbl>
## 1           46842
## 2           48023
## # ... with 1,289 more rows
```

```
# Select on columns names of the data frame which matches
# [ab] is interpreted as a regular expression ("a" or "b")
Ili.usa %>% select(matches("[ab]ge"))
```

```
## # A tibble: 1,291 x 6
##    age_0_4 age_25_49 age_25_64 age_5_24 age_50_64 age_65
##      <dbl>     <dbl>     <dbl>    <dbl>     <dbl>  <dbl>
## 1      179        NA       157      205        NA     29
## 2      199        NA       151      242        NA     23
## # ... with 1,289 more rows
```

```
# Select everything / all columns of the data frame
Ili.usa %>% select(everything())
```

```
## # A tibble: 1,291 x 16
##   region_type region    year  week weighted_ili
##   <chr>       <chr>    <int> <int>        <dbl>
## 1 National    National  1997    40         1.10
## 2 National    National  1997    41         1.20
## # ... with 1,289 more rows, and 11 more variables:
## #   unweighted_ili <dbl>, age_0_4 <dbl>,
## #   age_25_49 <dbl>, age_25_64 <dbl>, age_5_24 <dbl>,
## #   age_50_64 <dbl>, age_65 <dbl>, ilitotal <dbl>,
## #   num_of_providers <dbl>, total_patients <dbl>,
## #   week_start <date>
```

```
# Select all columns between year and ilitotal
Ili.usa %>% select(year:ilitotal)
```

```
## # A tibble: 1,291 x 11
##    year  week weighted_ili unweighted_ili age_0_4
##   <int> <int>        <dbl>          <dbl>   <dbl>
## 1  1997    40         1.10           1.22     179
## 2  1997    41         1.20           1.28     199
## # ... with 1,289 more rows, and 6 more variables:
## #   age_25_49 <dbl>, age_25_64 <dbl>, age_5_24 <dbl>,
## #   age_50_64 <dbl>, age_65 <dbl>, ilitotal <dbl>
```

```
# Select all columns except region_type and region.
Ili.usa %>% select(-c(region_type, region))
```

```
## # A tibble: 1,291 x 14
##    year  week weighted_ili unweighted_ili age_0_4
##   <int> <int>        <dbl>          <dbl>   <dbl>
## 1  1997    40         1.10           1.22     179
## 2  1997    41         1.20           1.28     199
## # ... with 1,289 more rows, and 9 more variables:
## #   age_25_49 <dbl>, age_25_64 <dbl>, age_5_24 <dbl>,
## #   age_50_64 <dbl>, age_65 <dbl>, ilitotal <dbl>,
## #   num_of_providers <dbl>, total_patients <dbl>,
## #   week_start <date>
```

```r
# Change variable names;
# num_range: Matches a numerical range like x01, x02, x03.
Ili.usa %>%
  rename(X1 = age_0_4, X2 = age_5_24, X3 = age_25_64, X4 = age_65) %>%
  select(num_range("X", 1:4))
```

```
## # A tibble: 1,291 x 4
##       X1    X2    X3    X4
##    <dbl> <dbl> <dbl> <dbl>
## 1    179   205   157    29
## 2    199   242   151    23
## # ... with 1,289 more rows
```

```r
# Select last variable, possibly with an offset.
Ili.usa %>% select(last_col())
```

```
## # A tibble: 1,291 x 1
##    week_start
##    <date>
## 1 1997-09-28
## 2 1997-10-05
## # ... with 1,289 more rows
```

2.3.5.2 More on Extracting Rows

Next, we will learn some useful functions for extracting rows, rather than using filter() and slice_*() functions in the previous section.

- distinct(): removes duplicate rows.
- sample_frac(): randomly selects fraction of rows.
- sample_n(): randomly selects n rows.
- slice(): selects rows by position.

```r
Ili.usa.2004 <- Ili.usa %>%
  dplyr::filter(year == 2004) %>%
  dplyr::select(week_start, weighted_ili)
```

```
Ili.usa.2005 <- Ili.usa %>%
  dplyr::filter(year == 2005) %>%
  dplyr::select(week_start, weighted_ili)

Ili.usa0405 <- Ili.usa.2004 %>% add_row(Ili.usa.2005)



# Remove duplicate rows.
Ili.usa0405 %>%
  add_row(Ili.usa.2005) %>%
  distinct()



## # A tibble: 104 x 2
##    week_start weighted_ili
##    <date>            <dbl>
## 1 2004-01-04         2.89
## 2 2004-01-11         2.10
## # ... with 102 more rows



# Randomly select n rows.
Ili.usa0405 %>% dplyr::sample_n(5)



## # A tibble: 5 x 2
##    week_start weighted_ili
##    <date>            <dbl>
## 1 2005-09-18         1.01
## 2 2004-09-19         0.692
## 3 2005-02-06         5.06
## 4 2004-07-18         0.495
## 5 2004-08-22         0.663



# Randomly select fraction of rows.
dplyr::sample_frac(Ili.usa0405, 0.5, replace = TRUE)



## # A tibble: 52 x 2
```

```
##    week_start weighted_ili
##    <date>             <dbl>
## 1 2005-09-11          0.855
## 2 2004-09-26          0.783
## # ... with 50 more rows
```

```
# Randomly select n rows.
dplyr::sample_n(Ili.usa0405, 10, replace = TRUE)
```

```
## # A tibble: 10 x 2
##    week_start weighted_ili
##    <date>             <dbl>
## 1 2004-06-06          0.835
## 2 2004-11-21          1.74
## # ... with 8 more rows
```

```
# Select rows by position.
dplyr::slice(Ili.usa0405, 10:15)
```

```
## # A tibble: 6 x 2
##    week_start weighted_ili
##    <date>             <dbl>
## 1 2004-03-07          1.03
## 2 2004-03-14          1.06
## 3 2004-03-21          0.849
## 4 2004-03-28          0.747
## 5 2004-04-04          0.662
## 6 2004-04-11          0.680
```

2.4   Making New Variables with `mutate()`

We will frequently need to add new columns depending on the values of current columns, such as to calculate the number of daily new cases based on the cumulative count. For this, we can use the `mutate()` function.

```
# Create a new variable Y2020.12.31 (new count 2020.12.31)
I.county.new <- I.county %>%
    dplyr::filter(State == "Florida") %>%
    dplyr::select(ID, County, X2020.12.31:X2020.12.30) %>%
    mutate(Y2020.12.31 = X2020.12.31 - X2020.12.30) %>%
    print(n = 2)
```

```
## # A tibble: 67 x 5
##        ID County   X2020.12.31 X2020.12.30 Y2020.12.31
##     <int> <chr>          <int>       <int>       <int>
## 1 12001 Alachua        15866       15674         192
## 2 12003 Baker           2470        2453          17
## # ... with 65 more rows
```

If we want to obtain the number of daily new cases based on the cumulative count for the dates in December only, we can try the following:

```
# Create variables Y2020.12.22 : Y2020.12.31
# with daily new count

FL.I.county <- I.county %>%
  dplyr::filter(State == "Florida")

I.county.tmp <- FL.I.county[, -(1:3)]
FL.I.county.new <- FL.I.county
FL.I.county.new[, -(1:3)] <- I.county.tmp -
  cbind(I.county.tmp[, -1], 0)

FL.I.county.DEC <- FL.I.county.new %>%
dplyr::select(ID, County, X2020.12.31:X2020.12.22)

name.tmp <- substring(names(FL.I.county.DEC)[-(1:2)], 2)
names(FL.I.county.DEC)[-(1:2)] <- paste0("Y", name.tmp)
FL.I.county.DEC %>%
    print(n = 2)
```

```
## # A tibble: 67 x 12
##        ID County   Y2020.12.31 Y2020.12.30 Y2020.12.29
##     <int> <chr>          <int>       <int>       <int>
## 1 12001 Alachua          192         100         101
## 2 12003 Baker             17          21          18
```

```
## # ... with 65 more rows, and 7 more variables:
## #   Y2020.12.28 <int>, Y2020.12.27 <int>,
## #   Y2020.12.26 <int>, Y2020.12.25 <int>,
## #   Y2020.12.24 <int>, Y2020.12.23 <int>,
## #   Y2020.12.22 <int>
```

2.5   Summarizing Data

Many data analysis tasks are able to be tackled while using the split-apply-combine paradigm. We can divide the data into groups, apply some analysis to each group separately, and then combine the results. The `group_by()` function in "dplyr" makes this very simple.

Various summary functions, which take a vector of values and return a single value, are used via the `summarize()` function; there are many types of summary functions as follows:

- `dplyr::first`: the first value of a vector;
- `dplyr::last`: the last value of a vector;
- `dplyr::nth`: the nth value of a vector;
- `dplyr::n`: the number of values in a vector;
- `dplyr::n_distinct`: the number of distinct values in a vector;
- `IQR`: the IQR of a vector;
- `min`: the minimum value in a vector;
- `max`: the maximum value in a vector;
- `mean`: the mean value of a vector;
- `median`: the median value of a vector;
- `var`: the variance of a vector;
- `sd`: the standard deviation of a vector.

The `group_by()` function is often paired with `summarize()`, which condenses each group into a single-row summary of that group. The `group_by()` takes the column as arguments that contain the categorical variables for which we want to calculate the summary statistics. Once the data are grouped, we can also summarize multiple variables simultaneously (and not necessarily on the same variable). So to compute the state-level cumulative infected count by State, we can try the following:

```
# State-level cumulative infected count
# Method 1: summarize()
I.state <- I.county %>%
  group_by(State) %>%
```

```
  summarize(across(X2020.12.31:X2020.01.22,
                 ~ sum(.x, na.rm = TRUE)))

# head(I.state, 2)[1:10]
I.state %>%
  print(n = 2, max_extra_cols = 5)
```

```
## # A tibble: 49 x 346
##    State    X2020.12.31 X2020.12.30 X2020.12.29 X2020.12.28
##    <chr>          <int>       <int>       <int>       <int>
## 1 Alabama       361226      356820      351804      347894
## 2 Arizona       523829      515366      510548      504616
## # ... with 47 more rows, and 341 more variables:
## #   X2020.12.27 <int>, X2020.12.26 <int>,
## #   X2020.12.25 <int>, X2020.12.24 <int>,
## #   X2020.12.23 <int>, ...
```

or we can use summarize_at(), which affects variables selected with a character
vector or vars():

```
# State-level cumulative infected count
# Method 2: summarize_at()
I.state <- I.county %>%
  group_by(State) %>%
  summarize_at(vars(X2020.12.31:X2020.01.22),
             ~ sum(.x, na.rm = TRUE))

# head(I.state, 2)[1:10]
I.state %>%
  print(n = 2, max_extra_cols = 5)
```

```
## # A tibble: 49 x 346
##    State    X2020.12.31 X2020.12.30 X2020.12.29 X2020.12.28
##    <chr>          <int>       <int>       <int>       <int>
## 1 Alabama       361226      356820      351804      347894
## 2 Arizona       523829      515366      510548      504616
## # ... with 47 more rows, and 341 more variables:
## #   X2020.12.27 <int>, X2020.12.26 <int>,
## #   X2020.12.25 <int>, X2020.12.24 <int>,
## #   X2020.12.23 <int>, ...
```

or we can use `summarize_if()`, which affects variables selected with a predicate function:

```
# State-level cumulative infected count
# Method 3: summarize_if()
I.state <- I.county %>%
  group_by(State) %>%
  summarize_if(is.numeric, ~ sum(.x, na.rm = TRUE))

# head(I.state, 2)[1:10]
I.state %>%
  print(n = 2, max_extra_cols = 5)
```

```
## # A tibble: 49 x 348
##    State       ID X2020.12.31 X2020.12.30 X2020.12.29
##    <chr>    <int>       <int>       <int>       <int>
## 1 Alabama  71489      361226      356820      351804
## 2 Arizona  60208      523829      515366      510548
## # ... with 47 more rows, and 343 more variables:
## #   X2020.12.28 <int>, X2020.12.27 <int>,
## #   X2020.12.26 <int>, X2020.12.25 <int>,
## #   X2020.12.24 <int>, ...
```

It can sometimes be useful to rearrange the result of a query to inspect the values. For example, we could sort on X2020.12.31 to put the group with the largest cumulative infected count first, using the `arrange()` function:

```
# State level cumulative infected count
# Method 4: sort by the cum. infected count
I.state <- I.county %>%
  group_by(State) %>%
  summarize_if(is.numeric, ~ sum(.x, na.rm = TRUE)) %>%
  arrange(desc(X2020.12.31))

# head(I.state, 2)[1:10]
I.state %>%
  print(n = 2, max_extra_cols = 5)
```

```
## # A tibble: 49 x 348
##    State            ID X2020.12.31 X2020.12.30 X2020.12.29
##    <chr>         <int>       <int>       <int>       <int>
```

```
## 1 California    351364     2307706     2275505     2242983
## 2 Texas       12256516     1772163     1755095     1735738
## # ... with 47 more rows, and 343 more variables:
## #    X2020.12.28 <int>, X2020.12.27 <int>,
## #    X2020.12.26 <int>, X2020.12.25 <int>,
## #    X2020.12.24 <int>, ...
```

In the above, `desc()` is used to re-order the data by a column in descending order.

2.6 Combining Datasets

R has multiple quick and sophisticated ways to join data frames by a common column. There are at least three ways:

- Base R's `merge()` function;
- Join family of functions from `dplyr`;
- Bracket syntax based on `data.table`.

2.6.1 The "Join" Family

The "dplyr" package uses SQL database syntax for the join family. For example, a **left join** means: include all the rows on the left and only those from the right data frame that match. For data frames `df1` and `df2`, if the join columns have the same name, all we need to do is `left_join(df1, df2)`. If they do not have the same name, we need a by argument, such as `left_join(df1, df2, by = c("df1ColName" = "df2ColName"))`. See an illustration in Figure 2.2.

Different join functions regulate what happens to rows that exist in one table but not the other.

- `left_join`: maintains all entries from the left (first) table and excludes any from the right table;
- `right_join`: keeps all entries from the right table and excludes any from the left table;
- `inner_join`: preserves just the items that are present in both tables;
- `full_join`: retains all of the entries in both tables, regardless of whether they occur in the other table.

An additional illustration can be found in Figure 2.3.

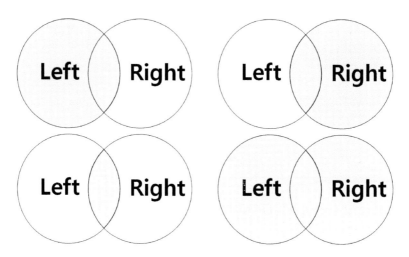

FIGURE 2.2: An illustration of left join (top left) and right join (top right), and inner join (bottom left) and full join (bottom right).

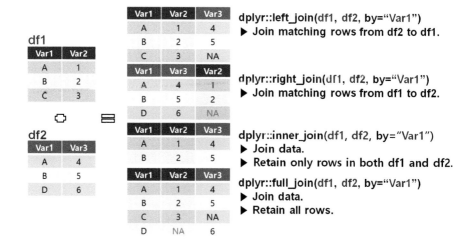

FIGURE 2.3: An illustration of the join functions.

2.6.2 Toy Examples with Joins

```
a <- tibble(x1 = LETTERS[c(1:3)], x2 = 1:3)
b <- tibble(x1 = LETTERS[c(1, 2, 4)], x3 = c(T, F, T))
a
```

```
## # A tibble: 3 x 2
##    x1          x2
##    <chr> <int>
## 1 A           1
## 2 B           2
## 3 C           3
```

```
b
```

```
## # A tibble: 3 x 2
##    x1         x3
##    <chr> <lgl>
## 1 A        TRUE
## 2 B        FALSE
## 3 D        TRUE
```

```
# Include all rows in a and b
inner_join(a, b, by = "x1")
```

```
## # A tibble: 2 x 3
##    x1          x2 x3
##    <chr> <int> <lgl>
## 1 A           1 TRUE
## 2 B           2 FALSE
```

```
# Return all rows from a
left_join(a, b, by = "x1")
```

```
## # A tibble: 3 x 3
##   x1       x2 x3
##   <chr> <int> <lgl>
## 1 A         1 TRUE
## 2 B         2 FALSE
## 3 C         3 NA
```

```
# Return all rows from b
right_join(a, b, by = "x1")
```

```
## # A tibble: 3 x 3
##   x1       x2 x3
##   <chr> <int> <lgl>
## 1 A         1 TRUE
## 2 B         2 FALSE
## 3 D        NA TRUE
```

```
# Include all rows in a or b
full_join(a, b, by = "x1")
```

```
## # A tibble: 4 x 3
##   x1       x2 x3
##   <chr> <int> <lgl>
## 1 A         1 TRUE
## 2 B         2 FALSE
## 3 C         3 NA
## 4 D        NA TRUE
```

```
# Include the rows in a that are not in b
anti_join(a, b, by = "x1")
```

```
## # A tibble: 1 x 2
##   x1       x2
##   <chr> <int>
## 1 C         3
```

```
# Want everything that doesn't match?
full_join(anti_join(a, b, by = "x1"),
          anti_join(b, a, by = "x1"), by = "x1")
```

```
## # A tibble: 2 x 3
##   x1        x2 x3
##   <chr> <int> <lgl>
## 1 C         3 NA
## 2 D        NA TRUE
```

2.6.3   Practicing with Joins for Real Data

Example 2.3. (US County-Level Population Data). We first get the data named pop.county from the GitHub R package IDDA. Note that there are four variables in this data: ID (county-level Federal Information Processing System code), County (name of county matched with ID), State (name of a state matched with ID), population (population of a county matched with ID).

```
data(I.county)
data(pop.county)
# Make I.county a tibble with as_tibble()
I.county <- as_tibble(I.county)
dim(I.county)
```

```
## [1] 3104   349
```

```
# Make pop.county a tibble with as_tibble()
pop.county <- as_tibble(pop.county)
dim(pop.county)
```

```
## [1] 3142    4
```

Now, we would like to join the two tables: I.county and pop.county using the left_join() as follows:

```
pop.county.tmp <- pop.county %>% dplyr::select(-c(County, State))
I.county.w.pop <- left_join(I.county, pop.county.tmp, by = "ID")
```

or we can:

```
I.county.w.pop <- left_join(I.county,
                    dplyr::select(pop.county, c(ID, population)),
                    by = "ID")
```

Example 2.4. (US State-level COVID-19 Data). In this example, we would like to create a map to show the risk of COVID-19 infection in each state of the US. First, we need to have a new dataset that contains infection risk and the geographic information of each state. We will get the infected count and state population in the state.long dataset in the IDDA package. state.long is a data frame with 16,905 rows and seven columns. Next, we obtain the boundary information of each state downloaded from PublicaMundi[1]. Then, we merge the two datasets to create a new dataset. We need the R package "geojsonio" to read the data from PublicaMundi.

```
library(geojsonio)
```

```
## Registered S3 method overwritten by 'geojsonsf':
##    method         from
##    print.geojson geojson
```

```
##
## Attaching package: 'geojsonio'
```

```
## The following object is masked from 'package:base':
##
##      pretty
```

---

[1] https://raw.githubusercontent.com/PublicaMundi/MappingAPI/master/data/geojson/us-states.json

```
library(IDDA)
data(state.long)
# Get the geospatial information from PublicaMundi
urlRemote  <- "https://raw.githubusercontent.com/"
pathGithub <- "PublicaMundi/MappingAPI/master/data/geojson/"
fileName   <- "us-states.json"
states0 <- geojson_read(x = paste0(urlRemote, pathGithub, fileName),
                        what = "sp")
head(states0@data)
```

```
##    id       name density
## 1 01    Alabama  94.650
## 2 02     Alaska   1.264
## 3 04    Arizona  57.050
## 4 05   Arkansas  56.430
## 5 06 California 241.700
## 6 08   Colorado  49.330
```

```
states1 <- states0
states1@data <- states0@data %>%
# Remove the space in the name of state if there is one
mutate(name_ns = sapply(name, gsub, pattern = " ", replacement = ""))
# The following merge step can be done using the join functions in
  ↳ dplyr
states1@data <- left_join(states1@data, state.long %>%
  filter(DATE == as.Date('2020-12-31')), by = c('name_ns' = 'State'))
# Calculate the risk of infection
states1@data <- states1@data %>%
  mutate(Infect_risk = Infected / pop)
```

2.6.4 More on Combining Rows of Tables

We discussed how to combine columns (variables) in the previous section. Here, we will see how to combine rows (cases) of datasets using the following set operations:

- `intersect(x,y)`: rows in both x and y;
- `union(x,y)`: rows in either or both x and y;
- `setdiff(x,y)`: rows that in x but not y;

- setequal(x,y): checks if two objects x and y are equal.

Let us look at the usage of the set operations with the ILI data.

```
Ili.usa2014_2015 <- Ili.usa %>% filter(year >= 2014, year <=2015)
Ili.usa2014 <- Ili.usa %>% filter(year >= 2014)
Ili.usa2016 <- Ili.usa %>% filter(year >= 2016)

# Rows that appear in both Ili.usa2014 and Ili.usa2016.
Ili.usa.intersect <- intersect(Ili.usa2014, Ili.usa2016)
Ili.usa.intersect %>%
  print(n = 2, max_extra_cols = 5)
```

```
## # A tibble: 338 x 16
##   region_type region    year  week weighted_ili
##   <chr>       <chr>    <int> <int>        <dbl>
## 1 National    National  2016     1         1.94
## 2 National    National  2016     2         2.00
## # ... with 336 more rows, and 11 more variables:
## #   unweighted_ili <dbl>, age_0_4 <dbl>,
## #   age_25_49 <dbl>, age_25_64 <dbl>, age_5_24 <dbl>,
## #   ...
```

```
# Rows that appear in either or both Ili.usa2014 and Ili.usa2016.
Ili.usa.union <- union(Ili.usa2014_2015, Ili.usa2016)
Ili.usa.union %>%
  print(n = 2, max_extra_cols = 5)
```

```
## # A tibble: 443 x 16
##   region_type region    year  week weighted_ili
##   <chr>       <chr>    <int> <int>        <dbl>
## 1 National    National  2014     1         4.28
## 2 National    National  2014     2         3.56
## # ... with 441 more rows, and 11 more variables:
## #   unweighted_ili <dbl>, age_0_4 <dbl>,
## #   age_25_49 <dbl>, age_25_64 <dbl>, age_5_24 <dbl>,
## #   ...
```

```
# Rows that appear in Ili.usa2014 but not Ili.usa2016.
Ili.usa.setdiff <- setdiff(Ili.usa2014, Ili.usa2016)
Ili.usa.setdiff %>%
  print(n = 2, max_extra_cols = 5)
```

```
## # A tibble: 105 x 16
##    region_type region     year  week weighted_ili
##    <chr>       <chr>     <int> <int>        <dbl>
## 1 National     National   2014     1         4.28
## 2 National     National   2014     2         3.56
## # ... with 103 more rows, and 11 more variables:
## #   unweighted_ili <dbl>, age_0_4 <dbl>,
## #   age_25_49 <dbl>, age_25_64 <dbl>, age_5_24 <dbl>,
## #   ...
```

```
# Check if two objects are equal.
setequal(Ili.usa2014_2015, Ili.usa2014)
```

```
## [1] FALSE
```

```
setequal(Ili.usa.union, Ili.usa2014)
```

```
## [1] TRUE
```

2.7 Data Reshaping

We may need to transform data from a wide to a long format. Many functions in R assume that data is in a long format rather than a wide format. Programs like SPSS, on the other hand, frequently employ data in a wide format. Let's consider the dataset I.state as an example. The column names "XYYYY.MM.DD" are not names of variables but values of a variable, which contains the values of the cumulative infected count. We need to pivot the column names into new variables.

Below we examine the pivot_longer() and pivot_wider() functions from the "tidyr" package.

2.7.1 From Wide to Long

We would like to change the data `I.state` from wide format to long format.

```
library(IDDA)
data(I.state)
names(I.state)
```

The function `pivot_longer()` is an updated version of the function `gather()`, which is no longer under active development. Compared to `gather()`, `pivot_longer()` is simpler to use and is able to handle more complex situations. Figure 2.4 demonstrates the idea of the `pivot_longer` function.

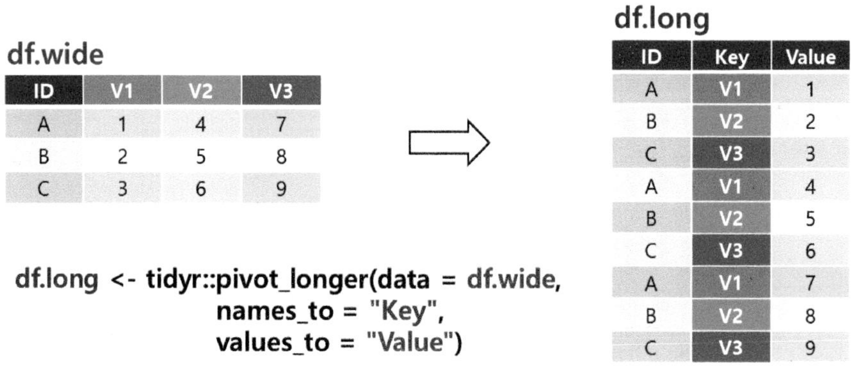

FIGURE 2.4: An illustration of the `pivot_longer()` function.

First, we take a look at the arguments of `pivot_longer()`:

- `data`: the data frame we want to morph from wide to long;
- `cols`: the columns that we need to pivot into longer format;
- `names_to`: the name of the new column that stores the column names of `data` or `cols` if specified;
- `values_to`: the name of the new column that stores all cell values of `data` or cell values from columns `cols` if specified.

```
I.state.wide <- I.state
dim(I.state.wide)
```

```
## [1]  49 347
```

```
I.state.long <- I.state.wide %>%
  # Select the columns to pivot
  pivot_longer(cols = X2020.12.31:X2020.01.21,
               names_to = "DATE",
               values_to = "Infected") %>%
  # Transform the DATE cell values to Date format
  mutate(DATE = as.Date(DATE, "X%Y.%m.%d"))

dim(I.state.long)
```

```
## [1] 16954     3
```

```
I.state.long %>%
  print(n = 2, max_extra_cols = 5)
```

```
## # A tibble: 16,954 x 3
##    State    DATE          Infected
##    <chr>    <date>           <int>
## 1 Alabama 2020-12-31      361226
## 2 Alabama 2020-12-30      356820
## # ... with 16,952 more rows
```

2.7.2 From Long to Wide

We can also transform the data back to a wide format. Indeed we can use the inverse transformation of `pivot_longer()`, the function `pivot_wider()` to "widen" the data, i.e., increasing the number of columns while decreasing the number of rows. The idea of the `pivot_wider()` function is illustrated in Figure 2.5.

The main arguments are as follows:

- data: the data frame we want to morph from wide to long;
- names_from: the column(s) whose cell values are used as the names of the new column(s);
- values_from: the column(s) whose cell values are used as the cell values of the new column(s).

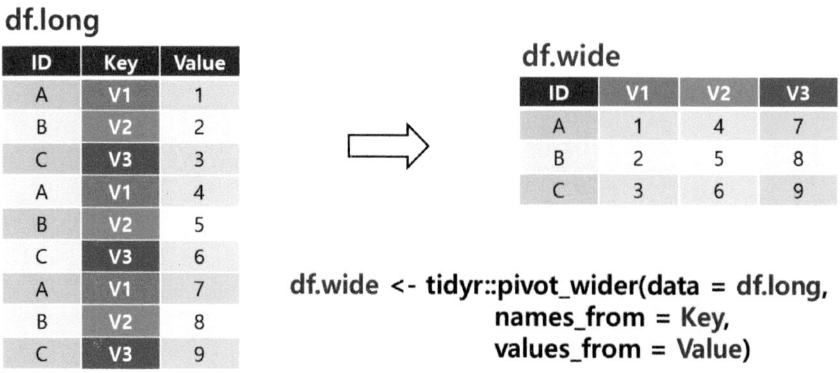

df.long

ID	Key	Value
A	V1	1
B	V2	2
C	V3	3
A	V1	4
B	V2	5
C	V3	6
A	V1	7
B	V2	8
C	V3	9

df.wide

ID	V1	V2	V3
A	1	4	7
B	2	5	8
C	3	6	9

**df.wide <- tidyr::pivot_wider(data = df.long,
names_from = Key,
values_from = Value)**

FIGURE 2.5: An illustration of the `pivot_wider()` function.

```
I.state.wide <- I.state.long %>%
    # Create new columns for `DATE`
    pivot_wider(names_from = DATE,
    # Use values from the `Infected` column
              values_from = Infected,
    # Add prefix to column names
              names_prefix = "X")

dim(I.state.wide)
```

```
## [1]   49 347
```

```
I.state.wide %>%
  print(n = 2, max_extra_cols = 5)
```

```
## # A tibble: 49 x 347
##    State  `X2020-12-31` `X2020-12-30` `X2020-12-29`
##    <chr>          <int>         <int>         <int>
## 1 Alabama       361226        356820        351804
## 2 Arizona       523829        515366        510548
## # ... with 47 more rows, and 343 more variables:
## #   X2020-12-28 <int>, X2020-12-27 <int>,
## #   X2020-12-26 <int>, X2020-12-25 <int>,
## #   X2020-12-24 <int>, ...
```

2.8 Further Reading

The major references used in this chapter are:

- Wickham, H. and Grolemund, G. (2016). *R for data science: Import, tidy, transform, visualize, and model data.* O'Reilly Media, Inc.
- Boehmke, B.C. (2016). *Data wrangling with R.* Springer International Publishing.

The following web resources are also helpful with regard to the topics covered in this chapter:

- `https://www.tidyverse.org/;`
- `https://dplyr.tidyverse.org/;`
- `https://tidyr.tidyverse.org/;`
- `https://r4ds.had.co.nz/wrangle-intro.html;`
- `https://datacarpentry.org/R-ecology-lesson/03-dplyr.html.`

2.9 Exercises

1. We are going to explore the basic data manipulation verbs of "dplyr" using `I.county`. Install the GitHub R package IDDA.

   ```
   library(IDDA)
   data(I.county)
   ```

   (a) Obtain a subset of the `I.county` by selecting `ID`, `County`, `State`.
   (b) Obtain a subset of the `I.county` by including all counties in California.
   (c) Obtain a subset of the `I.county` by including all counties that in the midwest states[2].

   ```
   Midwest = c("Illinois", "Michigan", "Indiana", "Ohio",
   "Wisconsin", "Florida", "Kansas", "Minnesota", "Missouri",
   "Nebraska", "SouthDakota", "NorthDakota")
   ```

---

[2]`https://www.census.gov/library/stories/state-by-state/midwest-region.html`

(d) Obtain a subset of the I.county by including the top ten counties from each midwest state based on the cumulative infected count on December 31, 2020.

(e) Obtain a subset of the I.county by including all the counties in California with the cumulative infected counts up to July 31, 2020.

(f) Create new columns of I.county by taking the logarithm of each count column.

(g) Sort the cumulative infected count on December 31, 2020 to find the state with the largest cumulative infected count.

2. Download the data pop.county from the GitHub R package IDDA.

(a) Join the tables of I.county and pop.county using inner_join, left_join , right_join, full_join. Do you get same or different tables?

(b) Based on the inner_join, create a table and name it I.pop.county, then create new columns of I.pop.county by dividing each count column by the population in the corresponding county, for example, risk.2020.12.31 = X.2020.12.31 / pop.

(c) For each state, list the top ten counties with the highest risk based on risk.2020.12.31.

3. Revisit the ilinet() function from the "cdcfluview" package to generate the "ILI" data.

(a) Create "ILI" data at the census regional level, and choose rows since the year 2015. Keep it in Ili.census2015.

(b) Randomly select 80% of rows from Ili.census without replacement and save them as Ili.usa.r1. Also, randomly select 1000 rows from Ili.census without replacement and save them as Ili.usa.r2.

(c) Using set operations, generate rows that appear in (1) both Ili.usa.r1 and Ili.usa.r2; (2) either or both Ili.usa.r1 and Ili.usa.r2; (3) Ili.usa.r1 but not Ili.usa.r2.

(d) Obtain subsets of data generated in part (c), including the Pacific region in 2020 with variable names containing the underline symbol ("_").

(e) Using three datasets in part (d), choose the last two columns (week_start and total_patients). Change the data to the wide format using week_start and total_patients as key and value, respectively.

3

Data Visualization with R Package "ggplot2"

There is a saying that a picture is worth a thousand words. Visualization reveals many underlying features of data, which statistics and models may miss: patterns, changes over time, unusual observations, clustering, gaps, relationships among variables. Thus, one can easily get intuition, effectively suggest interpretation, and eventually explore better models and forecasts by incorporating the features seen in graphs of the data. Although there are many base R graphics, the "ggplot2" R package has become more popular as a general scheme for data visualization due to its modern approach for multi-function displays with high quality.

This chapter introduces basic concepts and uses of visualization functions in the "ggplot2" R package. It examines the underlying grammar of graphics structure in the "ggplot2" package with examples. The primary focus is to understand how to apply those visualization techniques in "ggplot2" to real epidemiology data with various formats and purposes. After introducing the idea of individual and collective "geoms," we demonstrate two important collective plots: time series plots and maps. The chapter closes with a discussion of how to arrange plots and save the output.

3.1 An Introduction

The first thing to do is plot the data in the epidemiological data analysis task. Data visualization enables many features of the data to be displayed or summarized in a graphical format, including patterns, changes over time, unusual observations, relationships among variables, and spatial variations. The features seen in graphs of the data must then be incorporated as much as possible into the modeling or forecasting methods.

There are many types of graphs available, each with its own strengths and use cases. One of the challenges in the statistical learning process is choosing the appropriate visualization method to represent the data. Before constructing any display of epidemiological data, it is important to understand the type of task that we want to perform and determine the information to convey. Some common roles for data visualization include:

- highlighting a change from past patterns in the data;
- displaying a part-to-whole composition;
- showing how data is distributed;
- showing a difference or similarity between groups;
- displaying the spatial variation in geographical data;
- illustrating relationships among variables.

When the data is more complex, visualize broader patterns with graphs. Graphs can also visualize trends as well as identify variations from those trends. Variations in data may indicate substantial new results or just typographical or coding issues that need to be addressed. As a result, graphs may be useful tools for verifying and analyzing the data. Graphs can act as excellent visual aids for describing the data to others once the analysis is complete.

This chapter will describe the "ggplot2" introduced by Hadley (2016), and we will gain insight and practical skills for visualization of infectious disease data.

The package "ggplot2" builds on Leland Wilkinson's *The Grammar of Graphics* and focuses on the primacy of layers and adapts it for R.

> In brief, the grammar tells us that a graphic maps the data to the aesthetic attributes (colour, shape, size) of geometric objects (points, lines, bars). The plot may also include statistical transformations of the data and information about the plot's coordinate system. Facetting can be used to plot for different subsets of the data. The combination of these independent components are what make up a graphic. — Hadley (2016)

In the following, we will introduce the basics of "ggplot2" grammar and some of the key features.

3.2 Types of Variables and Preparation

Keep in mind that the primary purpose of preparing graphs is to communicate information. The types of variables we are analyzing and the media for the visualization can also affect our graphics practice.

3.2.1 Types of Variables

When examining data, we must know which data type we are working with in order to choose an appropriate display format. The data are most likely going to be in one of the following categories:

1. Categorical (Qualitative) variables

- A **nominal** variable is one whose values are categories without any numerical ranking. Good examples are occupation, place of birth, county of residence and diagnosis. The nominal variable is called **dichotomous** when it is characterized by only two classes. In epidemiology, it is common to see dichotomous variables: sex (male/female), exposure history (yes/no), alive or dead, ill or well, vaccinated or unvaccinated.

- An **ordinal** variable has values that can be ranked but are not necessarily evenly spaced. For example, the severity of illness may be categorized and ordered as "mild," "moderate" or "severe."

2. Numerical (Quantitative) variables

There are two types of **quantitative** variables:

- **Discrete** variables have values that are distinct and separate. Discrete data can't be measured but can be counted. For example, the number of new cases of a certain disease in a given year.

- **Continuous** variables represent measurements and can have any value in a range. An example of continuous data is the time from exposure to symptom onset. Discrete data can't be counted but can be measured.

Continuous variables can be further classified as either interval-scale or ratio-scale variables.

- An **interval-scale** variable is measured on a scale of equally spaced units but without a true zero point. An example of interval data is the date of birth.

- A **ratio-scale** variable is the same as interval values, with the difference that they do have an absolute zero. Good examples are height in centimeters or duration of illness.

3.2.2   Rules for Graph Designing

When designing graphs, we need to follow some rules to achieve the best practices, and Dicker and Gathany (1992) suggest the following:

- Check to ensure that a graphic can stand alone by clearly labeling the title, sources, axes, scales, and legends.

- Identify variables portrayed (legends or keys), which includes units of measure.

- Minimize the number of lines on a graph.

- In general, we depict frequency on the vertical scale, starting at zero, whereas we display the classification variable on the horizontal scale.

- Check to ensure that the scales for each axis are appropriate for the data presented.

- Define any acronyms or symbols.

- Specify any data that is not included.

3.2.3   Installing Packages and Loading Data

Before we begin, let's get ready by installing the "ggplot2" package by any of the following methods.

```
# The easiest way to get ggplot2 is to install the whole tidyverse:
install.packages("tidyverse")

# Alternatively, install just ggplot2:
install.packages("ggplot2")

# Or the development version from GitHub:
# install.packages("devtools")
devtools::install_github("tidyverse/ggplot2")
```

The package "devtools" downloads and installs the package from GitHub. By default, `install.packages()` is only able to install packages that are available in the Comprehensive R Archive Network (CRAN). In that case, the developer's tool, `devtools::install_github()`, enables users to install packages that have not been submitted to CRAN, but are available in GitHub.

In addition, we need to create a library of the required packages as follows.

In the rest of this section, we demonstrate how to create a basic scatterplot and output the figure in "png" and "rds" format. To create a "ggplot2" plot, we need to know three key components: (i) a data frame with each column being an attribute/variable, each row being an individual; (ii) a collection of aesthetic mappings between variables in the data and visual attributes; and (iii) at least one layer that explains how to portray each observation. A `geom_*()` function is commonly used to construct layers.

The "ggplot2" package generally prefers data in the "long" format: i.e., a column for every dimension, and a row for every observation. For illustration, we are going to use the `state.long` dataset in the R package IDDA. To prepare the data, install the R package IDDA from GitHub using the following command, which includes the datasets that we use for this book.

The `state.long` dataset includes the following variables: cumulative infected cases (`Infected`), cumulative death counts (`Death`), `Region`, `Division`, `State`, population (`pop`), and `DATE`, starting from January 22, 2020. Take a look at the first few lines using `head()`.

```
df <- IDDA::state.long
head(df)
```

```
## # A tibble: 6 x 7
##    State   Region Division        pop DATE        Infected
##    <chr>   <fct>  <fct>         <int> <date>         <int>
## 1 Alabama South  East South~ 4.89e6 2020-12-31      361226
## 2 Alabama South  East South~ 4.89e6 2020-12-30      356820
## 3 Alabama South  East South~ 4.89e6 2020-12-29      351804
## 4 Alabama South  East South~ 4.89e6 2020-12-28      347894
## 5 Alabama South  East South~ 4.89e6 2020-12-27      345623
## 6 Alabama South  East South~ 4.89e6 2020-12-26      343456
## # ... with 1 more variable: Death <int>
```

3.2.4   A Simple Scatterplot

Let's learn how to draw a simple scatterplot using the reported data of December 31, 2020. The variable `log(Infected)` is treated as the x-axis, and

log(Death) is treated as the y-axis. We create it by telling ggplot() the data
df, the aesthetic mapping aes(log(Infected + 1), log(Death + 1)), and
the layer geom_point(). The structure ggplot() + geom_point() is the typical
way to create a plot, in which ggplot() learns the data and mapping, and
geom_point() is a layer of a picture using the information embedded in gg-
plot(). Later in this chapter, we will show how to use "+" to assign additional
adjustments and add multiple layers to the existing figure. The following code
generates a scatterplot of the log cumulative death against the log cumulative
infected cases in the top left panel of Figure 3.1.

```
# Select the date
df <- IDDA::state.long %>%
  dplyr::filter(DATE == as.Date('2020-12-31'))

# Create scatterplot
# Data: df
# Aesthetic: first mapped to x, second mapped to y
# Layer: render the plot as a scatterplot
p <- ggplot(df, aes(log(Infected + 1), log(Death + 1)))
p + geom_point()
```

In the above example, the data frame df is the first parameter in the above
ggplot(), and aesthetics are defined within an aes() function. We need to
place + at the end of the previous line instead of the beginning of new line.

Aesthetics are plot features that may be used to highlight particular aspects
of the data. The following is a list of some common plot aesthetics we can
specify in the geom_point():

- x: position on the x-axis;
- y: position on the y-axis;
- alpha: transparency (1: opaque, 0: transparent);
- color: color of the border of elements;
- fill: color of the inside of elements;
- shape: shape;
- group: group;
- size: size;
- stroke: border size of points.

We will explain more details in the following sections.

3.3 Position Scales and Axes

3.3.1 Changing the Label of the Axis

We can convey information about the data by changing the labels of the axis using xlab() and ylab(). For example, we can change the label of the horizontal axis and vertical axis of the top left panel of Figure 3.1 to "log Infected" and "log Death." See the top right panel of Figure 3.1 for the customized labels and title.

```
# Change the transparency using alpha
p <- p + geom_point(alpha = 0.7) +
  # Change the label of horizontal axis
  xlab('log Infected') +
  # Change the label of vertical axis
  ylab('log Death') +
  # Change the title
  labs(title = 'Log death against infected cases in US')
p
```

3.3.2 Changing the Range of the Axis

For continuous variables, we can provide the lower and upper limits using xlim() and ylim(). For categorical variables, we can provide the names of categories desired. To suppress the warning "Removed XXX rows containing missing values," use na.rm. This needs to be carefully used because the data outside the range are converted to NA, which will affect later manipulations, such as calculating the mean or sum. See the bottom panel of Figure 3.1 for a scatterplot with a customized axis range for continuous and discrete features, respectively.

```
# For continuous variable, provide the lower and upper limits
p <- p + geom_point() + ylim(4, 12)
p
```

```
# For discrete variable, provide the names of categories desired
# To suppress the warning 'Removed XXX rows containing...', use
↪   `na.rm`.
df <- IDDA::state.long %>%
  dplyr::filter(DATE == as.Date('2020-12-31'))

ggplot(df, aes(Region, log(Death + 1))) +
  geom_point(na.rm = TRUE) +
  xlim('West', 'Midwest', 'South')
```

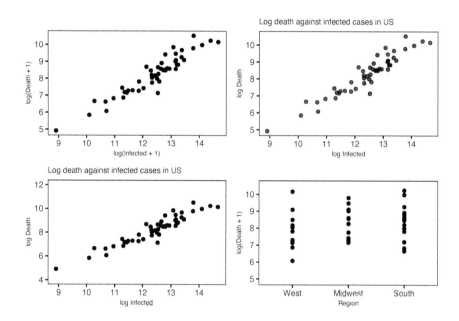

FIGURE 3.1: Various scatterplots of log cumulative death against log cumulative infected cases. Top left: base plot. Top right: customized labels and title. Bottom left: customized axis range for continuous features. Bottom right: customized axis range for discrete features.

3.4 Color Scales and Size of `geom_point()`

There are two ways of coloring. One approach is to color all points with the same color. The other method is to color the points according to a particular feature of the observation.

3.4.1 Changing the Color of All Points

To change the color of all points in a scatterplot, we can use the `color` argument inside `geom_point()`.

```
p + geom_point(color = "blue")
```

3.4.2 Coloring Observations by the Value of a Feature

We can map the colors of the observations to the class variable to demonstrate the class of each observation. The following example shows how to map the variable `Region` or `pop` to the color aesthetic, as given in the top and bottom panels of Figure 3.2, respectively.

```
# Use Region as the feature for coloring
p + geom_point(aes(color = Region))
# p + aes(color = Region)

# Use population for coloring
p + geom_point(aes(color = pop))
```

The `color` feature is located at different layers in the three figures. In the first figure, it is under `geom_point()`, while in the latter two figures, it is under `geom_point(aes())`. Because we only have one layer in this example, we can equivalently use `aes()`, i.e., the aesthetic mapping for the entire scatterplot.

3.4.3 Changing the Color Palette

In addition, we can personalize the color palette using `scale_fill_brewer()` for a discrete color scale and `scale_fill_distiller()` for a continuous color scale, as illustrated in the top panel of Figure 3.3.

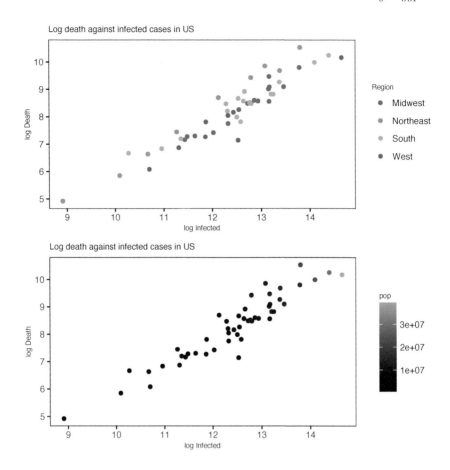

FIGURE 3.2: Top: a scatterplot with points colored by `Region`. Bottom: a scatterplot with points colored by `pop`.

```
# Change the palette
# For discrete scale
p + geom_point(aes(color = Region)) +
   scale_fill_brewer(palette = "Set1", aesthetics = "color")
# For continuous scale
p + geom_point(aes(color = pop)) +
  scale_fill_distiller(palette = 2, aesthetics = "color")
```

3.4.4 Changing the Size by the Value of a Feature

Similar to changing the color of the data points in the previous scatterplot, we can also control the point sizes by using the `size` argument inside `geom_point()` with `aes`; see the middle panel of Figure 3.3.

```
p <- ggplot(df, aes(log(Infected + 1), log(Death + 1))) +
  xlab('log Infected') +
  ylab('log Death') +
  labs(title = 'Log death against infected cases in US')

# Change the point size
p + geom_point(aes (size = pop))

# Change the point color and size
p + geom_point(aes (size = pop, color = pop))
```

We can also combine size and color legends into one; however, this only works if they are compatible; that is, they need to have exactly the same breaks. The following example demonstrates two ways of combining. The resulting plots are given by the bottom panel of Figure 3.3.

```
# Combine the color and size in legend
# Method 1: keep the size and color the same limits and breaks
p + geom_point(aes (size = pop, color = pop)) +
  scale_color_continuous(limits = c(0e7, 4e7),
                         breaks = seq(0, 4e7, by = 1e7)) +
  scale_size_area(limits = c(0e7, 4e7),
                  breaks = seq(0, 4e7, by = 1e7), max_size = 12) +
  guides(color = guide_legend(), size = guide_legend())

# Method 2: use scale_color_gradient and scale_size
p <- p + geom_point(aes(size = pop, color = pop), alpha = 0.7) +
  scale_color_gradient(low = "lightblue", high = "red") +
  scale_size_area(max_size = 12) +
  guides(color = guide_legend(), size = guide_legend())

p
```

For Method 1, the key to combining two aesthetic settings of the layer to one legend, in this case, `color` and `size`, is to set the `limits` and breaks to be the same in `guides()`. For both methods, `guides(color = guide_legend(), size = guide_legend())` is needed.

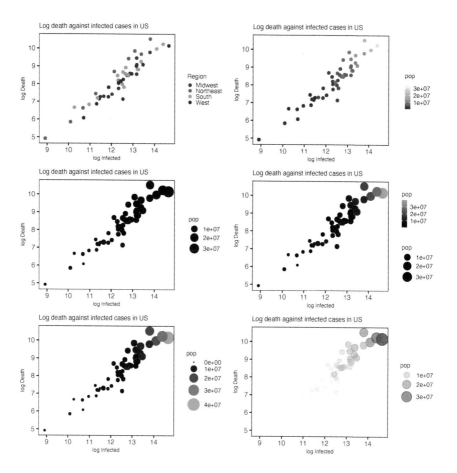

FIGURE 3.3: Various scatterplots. Top left: customized color palette for discrete scale. Top right: customized color palette for continuous scale. Middle left: customized point size. Middle right: customized point size and color. Bottom left: customized point size using Method 1. Bottom right: customized point size using Method 2.

3.4.5   An Example of a Row-labeled Dot Plot

Sometimes, it is interesting to compare one region to another in infectious disease studies. Taking COVID-19 studies as an example, we might be interested in assessing one state's current risk level compared to other states. It is well known that we can't directly compare one state to another purely based on the number of confirmed cases due to different population sizes. One statistic often used in the spatial representation of aggregated (e.g., county or state-level) disease risk data is the weekly local risk, defined as the average number of new cases in the past week divided by the population in the area. The following example demonstrates how to draw a row-labeled dot plot of weekly local risk for the 48 adjoining US states and the District of Columbia. We first load the data I.state and pop.state from the IDDA package, then merge the two datasets and compute the weekly local risk for each state from May 24 to May 31, 2020.

```
# Load the datasets I.state and pop.state
library(IDDA)
data(I.state)
data(pop.state)

# Compute the weekly new cases from May 24 to May 31, 2020
I.state.new <- I.state %>%
  mutate(X.W22 = X2020.05.31 - X2020.05.24) %>%
  dplyr::select(State, X.W22)

# Compute the weekly local risk
I.state.W22 <- inner_join(I.state.new, pop.state, by = "State") %>%
  mutate(X.W22.R = X.W22 / 7 / population * 100000)
head(I.state.W22)
```

```
## # A tibble: 6 x 4
##     State       X.W22 population X.W22.R
##     <chr>       <int>     <int>   <dbl>
## 1 Alabama      3475   4887871   10.2
## 2 Arizona      3597   7171646    7.17
## 3 Arkansas     1294   3013825    6.13
## 4 California  18367  39557045    6.63
## 5 Colorado     2325   5695564    5.83
## 6 Connecticut  1712   3572665    6.85
```

Figure 3.4 shows a state-labeled dot plot sorted by the average number of daily new cases per 100,000 people from May 24 to May 31, 2020.

```
library(ggplot2)
ggplot(I.state.W22, aes(x = X.W22.R,
                        y = reorder(State, X.W22.R))) +
```

```
geom_point(shape = 21, fill = "red", size = 3, color = "black") +
labs(x = "Cases per 100,000 Population",
    y = "State",
    title = "Average daily new cases per 100,000 people (May 24 to
    ↪  May 31)")
```

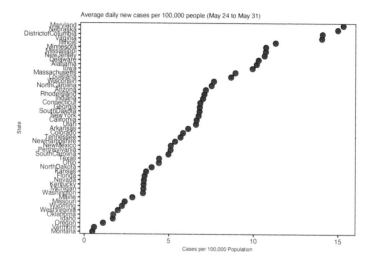

FIGURE 3.4: A state-labeled dot plot with an average number of daily new cases per 100,000 people from May 24 to May 31, 2020.

The `reorder()` function is used in the above code to put the rows (states) in top-down decreasing order according to the weekly local risk. We also change the default `geom_point()` aesthetic constants for size, color and shape, and choose the red color to fill the dots.

3.5 Individual Geoms

Apart from the scatterplot, there are many individual "geoms," for example:

- `geom_line()`: line graphs;
- `geom_boxplot()`:boxplots;
- `geom_bar()`: bar chart;
- `geom_histogram()`: histogram plots;
- `geom_smooth()`: regression lines or curves.

We introduce a few of them in detail as follows.

3.5.1 Histograms

The histogram is an important tool to summarize the range and frequency of observations, which provides an approximate representation of the distribution of numerical data. Here we plot the histogram of log daily new infected case counts using geom_histogram(). We can adjust the option binwidth to control the widths of the bins. Figure 3.5 shows examples using geom_histogram() based on the code below.

```r
# Prepare the daily new Infected for each state
# in the period 2020-11-12 to 2020-12-31
df <- IDDA::state.long %>%
  dplyr::filter(DATE <= '2020-12-31' & DATE > '2020-11-11') %>%
  group_by(State) %>% # Group by State
  mutate(Y.Infected =
           c(Infected[-length(Infected)] - Infected[-1], 0))
head(df)
```

```
## # A tibble: 6 x 8
## # Groups:   State [1]
##    State   Region Division        pop DATE       Infected
##    <chr>   <fct>  <fct>         <int> <date>        <int>
## 1 Alabama South  East South~ 4.89e6 2020-12-31     361226
## 2 Alabama South  East South~ 4.89e6 2020-12-30     356820
## 3 Alabama South  East South~ 4.89e6 2020-12-29     351804
## 4 Alabama South  East South~ 4.89e6 2020-12-28     347894
## 5 Alabama South  East South~ 4.89e6 2020-12-27     345623
## 6 Alabama South  East South~ 4.89e6 2020-12-26     343456
## # ... with 2 more variables: Death <int>,
## #   Y.Infected <dbl>
```

```r
p <- ggplot(df, aes(log(Y.Infected + 1)))
p + geom_histogram(binwidth = 1)
p + geom_histogram(binwidth = 1) + aes(fill = Region)
```

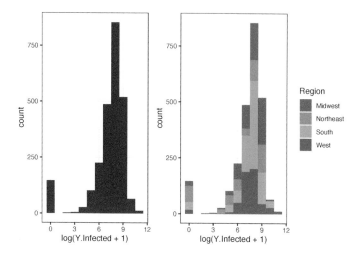

FIGURE 3.5: Histogram examples using `geom_histogram()`.

3.5.2   Bar Charts

The discrete analogue of a histogram is the bar chart that presents categorical data with rectangular bars.

3.5.3   The Default Bar Chart

The default `geom_bar()`, or equivalently `geom_bar(stat='count')`, counts the number of observations in each category shown as the top left panel of Figure 3.6. This plot essentially tells us how many states there are in each region.

```
df <- IDDA::state.long %>%
  dplyr::filter(DATE == as.Date('2020-12-31'))
p <- ggplot(df, aes(Region))
p + geom_bar()
```

3.5.4   Bar Charts with Assigned Values

In addition to the previous example, we can assign the height of the bars by ourselves by using the option `geom_bar(stat = 'identity')`. In that case, we tell `geom_bar` to use the y value in the data frame as the height of the bars. For those plots, see the top right and middle left panels of Figure 3.6.

```
df <- IDDA::state.long %>%
  dplyr::filter(DATE == as.Date('2020-12-31'))
p <- ggplot(df, aes(Region, Infected))
p + geom_bar(stat = 'identity')
p + geom_bar(stat = 'identity', aes(fill = Division))
```

3.5.5 Legends

The "ggplot2" package makes changing the location and the look of a graph legend simple.

3.5.5.1 Legend Positions

We can adjust the position of the legends using `theme(legend.position = 'left/right/bottom/none')`. The middle right panel of Figure 3.6 shows that the position of the legend is moved from the right to the bottom.

```
p <- ggplot(df, aes(Region, fill = Region)) +
  ylab('Number of states') +
  geom_bar()

p + theme(legend.position = 'bottom')
```

Note that the bar plot automatically explains the color of the regions, so the legend here is redundant. We can also remove the legend in a graph using `theme(legend.position = "none")`.

3.5.5.2 Legend Guide `guide_legend()`

We can also assign individual keys to the legend using the options of `guide_legend()`. Here we introduce some of the most useful options.

- `nrow` and `ncol`: specify the dimensions of the table.

- `byrow`: fills the rows, set to `FALSE` by default.

```
p + guides(fill = guide_legend(ncol = 2, byrow = TRUE))
```

- reverse: reverse the order of the keys.

```
p + guides(fill = guide_legend(reverse = TRUE))
```

Those plots using guide_legend(ncol = 2, byrow = TRUE) and guide_legend(reverse = TRUE) are given by the bottom left and right panels of Figure 3.6, respectively.

3.5.6   Boxplots, Jittering and Violin Plots

Conditioning on a categorical feature, or conditioning on groups, we may want to conduct a side-by-side comparison for a certain variable. We can use the following tools.

- Jittering: the function geom_jitter() adds a small amount of random noise to the data, which can help prevent over-plotting.
- Boxplots: the function geom_boxplot() summarizes the shape of the distribution with summary statistics.
- Violin plots: the function geom_violin() shows a compact representation of the "density" of the distribution and highlights the areas where more points are found.

The examples for a point, jittering, boxplot, and violin plots are illustrated in Figure 3.7.

```
df <- IDDA::state.long %>%
  dplyr::filter(DATE == as.Date('2020-12-31')) %>%
  mutate(Risk = Infected / pop)

p <- ggplot(df, aes(Region, Risk, color = Region))
p + geom_point()
p + geom_jitter()
p + geom_boxplot()
p + geom_violin()
```

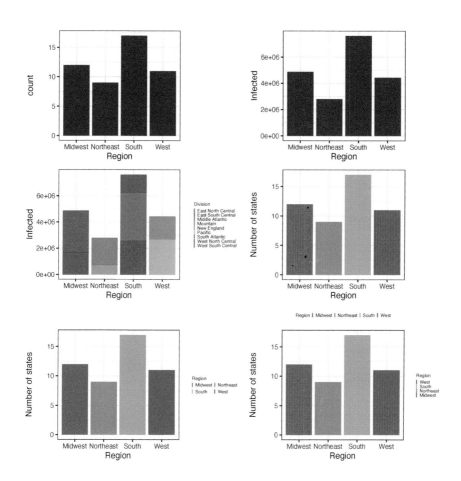

FIGURE 3.6: Various bar plots. Top left: a bar plot of the number of states in each region with geom_bar(). Top right: a bar plot of the number of infected cases in each region. Middle left: a bar plot of the number of infected cases with a legend at right. Middle right: a bar plot of the number of states in each region with a legend at the bottom. Bottom left: a bar plot of the number of states in each region using guide_legend(ncol = 2, byrow = TRUE). Bottom right: a bar plot of the number of states in each region using guide_legend(reverse = TRUE).

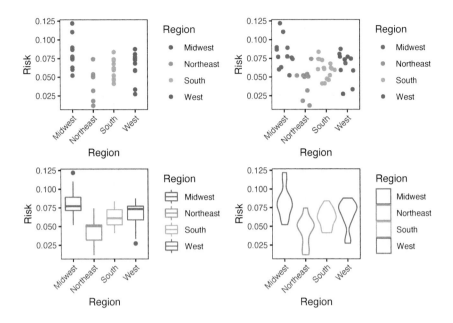

FIGURE 3.7: Top left: points plot. Top right: jittering plot. Bottom left: boxplot plot. Bottom right: violin plot.

3.6 Collective Geoms

For each observation (row), an individual "geom" can generate a unique graphical object. For example, the "point geom" draws one point for each row.

Several "geoms" may be added to the same ggplot object, which will allow us to build layers to design complex graphs and displays multiple observations with one geometric object. For example, we have previously created a scatterplot, and then we can add regressed lines on the top of the scatterplot layer. We can add more information from a statistical summary, or add a text geom to annotate the plot.

3.6.1 Smoothers

Sometimes it can be difficult to see the pattern on a scatterplot. A smoothing line can be helpful to identify what the pattern or trend looks like. On top of the scatterplot, we can add smoothed conditional means or regressed lines and a prediction band to it using geom_smooth().

A commonly used smoothing method is the LOcally WEighted Scatter plot Smoothing (or lowess) introduced by Cleveland (1979). It essentially fits local polynomial regressions and joins them together. For example, to fit at point x, the estimate is made using points in a neighborhood of x, weighted by their distance from x. One of the attractive features of this method is that no global function of any form is needed to fit a model to the data, thus allowing the data to speak for themselves.

By default, `geom_smooth()` adds a "loess" (locally estimated scatterplot smoothing) smoother to the data. We can call the layer either by `geom_smooth()` or `geom_smooth(method ='loess')`. In addition, we can adjust the option `span` to control the smoothness of the line. The higher `span` is, the less wiggly the line will be.

```
df <- IDDA::state.long %>%
  dplyr::filter(DATE == as.Date('2020-12-31'))
p <- ggplot(df, aes(log(Infected + 1), log(Death + 1))) +
  geom_point()
# Use `span` to control the smoothness of the line
# The higher `span` is, the less wiggle the line is
p + geom_smooth(method = 'loess', span = 0.5)
p + geom_smooth(method = 'loess', span = 1)
```

We can also use `method = 'lm'` to fit a simple linear model. To draw a polynomial of degree n, we can change the formula to `y ~ poly(x, n)`. Here is an example fitting a linear or quadratic regression line.

```
# Add a linear trend
p + geom_smooth(method = 'lm')
# Add a quadratic trend
p + geom_smooth(method = "lm", formula = y ~ poly(x, 2))
```

The plots using a smoothing method in the codes above are shown in Figure 3.8.

3.7 Time Series

Time series plots are very popular in the surveillance of infectious diseases.

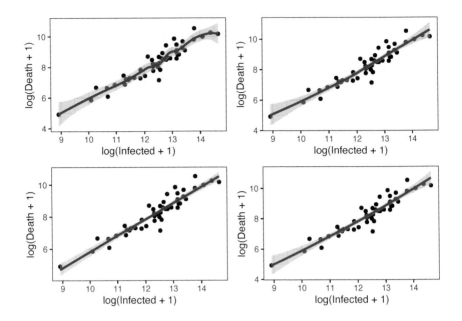

FIGURE 3.8: Top left: `loess` smoother example with `span = 0.5`. Top right: `loess` smoother example with `span = 1.0`. Bottom left: `lm` smoother example with a simple linear regression. Bottom right: `lm` smoother example with a quadratic regression.

3.7.1 Basic Line Plots

In traditional time series plots, we use time as the x-axis variable, and plot the time series using `geom_line()`.

In surveillance, time series of counts are often recorded. Below, we consider the time series of the daily new infected count for Iowa, US.

```
df <- IDDA::state.long %>%
  dplyr::filter(State == "Iowa") %>%
  arrange(DATE) %>%
  mutate(Y.Infected = Infected - lag(Infected)) %>%
  dplyr::filter(!is.na(Y.Infected))
```

To visualize the data, we draw a time series plot first, as in the top left panel of Figure 3.9.

```
p <- ggplot(df, aes(DATE, Y.Infected)) +
  geom_line() +
  labs(x = "Days", y = "Count",
       title = 'Daily new infected cases in Iowa') +
  theme(legend.position = 'bottom')
p
```

3.7.2 Adding a Second Line

Next, we display the prediction results on the time series plot in the top right panel of Figure 3.9. The prediction (`mean`) and prediction intervals (`lower` and `upper`) for the next 14 days are saved in the dataset: `IDDA::fore`.

```
df.pred <- as.data.frame(IDDA::fore[c('mean', 'lower', 'upper')])
names(df.pred) <- c('mean', 'lower', 'upper')
df.pred$DATE <- tail(df$DATE,1) +
  c(1:length(IDDA::fore$mean))
```

```
# Add a line for predictions
p + geom_line(mapping = aes(x = DATE,
                            y = mean,
                            color = 'Predicted Value'),
              linetype = "dashed",
              # Set the line type in legend
              key_glyph = "timeseries",
              data = df.pred) +
  scale_color_manual("", values = "red")
```

3.7.3 Adding Ribbons

Next, we show the prediction intervals. On top of the line plots, we can add another layer to the existing line, and create a line with two parts; see the middle left panel of Figure 3.9.

```
# Add prediction intervals
p <- p +
  geom_ribbon(mapping = aes(x = DATE,
                            y = mean,
                            ymin = lower,
                            ymax = upper,
                            fill = '95% Prediction Intervals'),
              data = df.pred, alpha = 0.2) +
# Add line for predicted value
geom_line(mapping = aes(x = DATE, y = mean,
                        color = 'Predicted Value'),
          linetype = "dashed", data = df.pred,
          # Set the line type in legend
          key_glyph = "timeseries") +
scale_color_manual("", values = "red")+
scale_fill_manual("", values = "pink")

p
```

It is worth noting that the layer added later is put on top; therefore, it is important to keep track of the order of the layers we add.

3.7.4   Adjusting the Scale of the Time Axis

There are multiple ways to define the ticks on the axis of dates and times. The labeled **major breaks** and further the **minor breaks** are not labeled but marked by grid lines. The options date_breaks and date_minor_breaks, respectively, can customize those breaks. The middle right panel of Figure 3.9 presents an example for adjusting the scale of the time axis from the code below.

```
# Adjust the scale of time axis
p + scale_x_date(
  limits = as.Date(c("2020-10-01", "2021-01-15")),
  date_breaks = "1 month",
  date_minor_breaks = "1 week",
  date_labels = "%B %Y"
)
```

In the above syntax, date_labels are set to a string of formatting codes that defines the order, format, and elements to be displayed in the above syntax:

- %d: day of the month (01-31);
- %m: month, numeric (01-12);
- %b: month, abbreviated (Jan-Dec);
- %B: month, full (January-December);
- %y: year, without century (00-99);
- %Y: year, with century (0000-9999).

3.7.5 Adding Annotations

It is often necessary to make annotations to the data displayed when constructing a data visualization. An annotation provides additional information for the data that is being displayed. Adding text to a plot, for example, is one of the most prevalent types of annotation. The function geom_text(), which adds label text at the supplied x and y locations, is the most used tool for labeling graphs. We can also add reference lines to the plot using geom_vline or geom_hline. The bottom left panel of Figure 3.9 shows an annotated time series plot with shades and reference lines for each epiweek.

```
library(lubridate)
# Prepare the data
df <- df %>%
  filter(DATE >= as.Date('2020-11-15')) %>%
  mutate(start = floor_date(DATE, "week")) %>%
  mutate(end = ceiling_date(DATE, "week")) %>%
  mutate(epiweeks = as.factor(epiweek(DATE)))

df.epiweeks <- df %>%
  dplyr::select(start, end, epiweeks) %>%
  unique()

# Draw the base ggplot
ggplot(df, aes(DATE, Y.Infected)) +
  labs(x = "Days", y = "Count",
       title = 'Daily new infected cases in Iowa') +

  # Add rectangle for each quarter
  geom_rect(
    aes(xmin = (start), xmax = (end), fill = epiweeks),
    inherit.aes = F, ymin = -Inf, ymax = Inf,
    alpha = 0.5, data = df.epiweeks) +
  scale_fill_brewer(palette = "Blues", aesthetics = "fill") +

  # Add a vertical line
```

```
geom_vline(aes(xintercept = as.numeric(start)),
  data = df, color = "gray",
  linetype = 'dashed', size = 0.5) +

# Add text
geom_text(
  aes(x = start, y = 0 , label = paste0('Week:', epiweeks)),
  data = df.epiweeks, inherit.aes = F,
  size = 2.5, vjust = 0, hjust = 0, nudge_x = 0.5) +

# Add time series lines
geom_line() +
geom_line(mapping = aes(x = DATE, y = mean,
                        color = 'Predicted Value'),
          linetype = "dashed", data = df.pred ,
          key_glyph = "timeseries") +
geom_segment(data = df,
             aes(x = max(DATE),
                 xend = min(df.pred$DATE),
                 y = Y.Infected[which.max(DATE)],
                 yend = df.pred$mean[which.min(df.pred$DATE)])) +
theme(legend.position = 'bottom')
```

3.8 Maps

In epidemiology, data often include geographical information such as latitude and longitude or regions like country, state, or county. To plot these types of data, we can extend an existing visualization onto a map background. We will learn how to make choropleth maps, sometimes called heat maps, using the "ggplot2" package. A **choropleth map** depicts a geographic landscape with units such as nations, states, or watersheds, each of which is colored according to a particular value.

We will illustrate the key steps of constructing a choropleth map using the "ggplot2" and the "maps" R packages; the latter contains a lot of outlines of continents, countries, states, and counties. In the example below, we are interested in drawing a US state-level map of COVID-19 related informa-tion. We first load necessary packages, and load the US state map data using `map_data("state")`. Then, we merge the spatial information with the `IDDA::state.long` dataset.

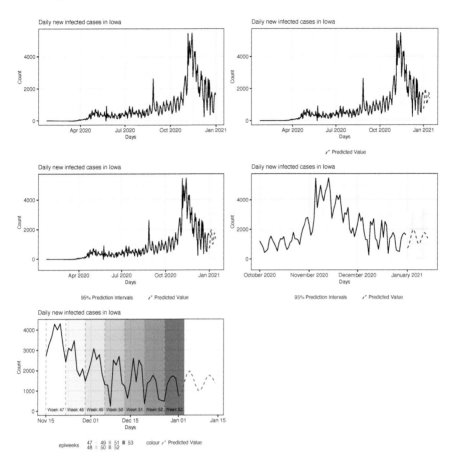

FIGURE 3.9: Various time series plots. Top left: basic plot. Top right: with added predictions. Middle left: with ribbons and second line. Middle right: with adjusted time range and format. Bottom left: annotated plot with shades and reference lines for each epiweek.

```
# Read map and data
library(ggplot2)
library(maps)
library(dplyr)
```

```
##
## Attaching package: 'maps'
```

```
## The following object is masked from 'package:purrr':
##
##      map
```

```
# Load the US state map data
MainStates <- map_data("state")
head(MainStates, 3)
```

```
##      long   lat group order  region subregion
## 1 -87.46 30.39     1     1 alabama      <NA>
## 2 -87.48 30.37     1     2 alabama      <NA>
## 3 -87.53 30.37     1     3 alabama      <NA>
```

```
MainStates <- MainStates %>%
  mutate('state' = gsub(' ', '', MainStates$region)) %>%
  select(-c('region', 'subregion'))
head(MainStates, 3)
```

```
##      long   lat group order   state
## 1 -87.46 30.39     1     1 alabama
## 2 -87.48 30.37     1     2 alabama
## 3 -87.53 30.37     1     3 alabama
```

```
state.long.shape <- IDDA::state.long %>%
  mutate('state' = tolower(IDDA::state.long$State)) %>%
  right_join(MainStates, by = 'state') %>%
  select(-state)
head(state.long.shape)
```

```
## # A tibble: 6 x 11
##   State   Region Division       pop DATE     Infected
##   <chr>   <fct>  <fct>        <int> <date>      <int>
## 1 Alabama South  East South~ 4.89e6 2020-12-31 361226
## 2 Alabama South  East South~ 4.89e6 2020-12-31 361226
## 3 Alabama South  East South~ 4.89e6 2020-12-31 361226
```

```
## 4 Alabama South   East South~ 4.89e6 2020-12-31    361226
## 5 Alabama South   East South~ 4.89e6 2020-12-31    361226
## 6 Alabama South   East South~ 4.89e6 2020-12-31    361226
## # ... with 5 more variables: Death <int>, long <dbl>,
## #   lat <dbl>, group <dbl>, order <int>
```

```
df <- state.long.shape %>%
  dplyr::filter(DATE == '2020-12-31') %>%
  mutate (Risk = Infected / pop * 1000)
```

3.8.1   Making a Base Map

Using qplot(), we can obtain our first map like the top left panel of Figure 3.10 using the following code:

```
qplot(long, lat, geom = "point", data = df)
```

We can use the geom_polygon() function to create a map with black borders and add light blue to fill in the map; see the top right panel of Figure 3.10.

```
# Plot all states with ggplot2, black borders and light blue fill
ggplot() +
  geom_polygon(data = df,
                aes(x = long, y = lat, group = group),
                color = "black", fill = "lightblue")
```

3.8.2   Customizing Choropleth Maps

Now that we have created a base map of the mainland states, we will color each state according to its risk as given in the middle left panel of Figure 3.10. Let us consider the IDDA::ggplot_map_state dataset.

```
# Create a choropleth map of the US
p <- ggplot() + geom_polygon(data = df,
          aes(x = long, y = lat, group = group,
              fill = Risk),
          color = "white", size = 0.2) +
  theme(legend.position = 'bottom')
p
```

In the above example, each state is colored by "Infected per 1000 people" to make the legend easier to read. Border color (white) and line thickness (size = 0.2) are specifically defined within this geom_polygon().

It is often helpful to adjust color schemes, figure out how to deal with missing values (na.values), and formalize labels after a map has been constructed. It is worth noting that we gave the graph a name, p. This will come in handy when we add more components to the map.

The example for the plot can be found in the middle right panel of Figure 3.10 based on the code below.

```
p + scale_fill_continuous(name = "Infected per 1000 pop",
                          low = "yellow", high = "darkred",
                          limits = c(0, 125),
                          breaks = c(5, 25, 50, 75, 100, 125),
                          na.value = "grey50") +
  labs(title = "Infected per 1000 population on 2020-12-31")
```

3.8.3 Overlaying Polygon Maps

It is also feasible to overlay two polygon maps. The code below adds a thicker line to illustrate state boundaries after creating county borders with a small line size. The alpha = .3 makes the fill in the state map transparent, enabling us to view the county map behind it; see the bottom left panel of Figure 3.10.

```
ggplot() + geom_polygon(data = map_data("county"),
            aes(x = long, y = lat, group = group),
            color = "darkblue",
            fill = "lightblue", size = .1) +

         geom_polygon(data = map_data('state'),
               aes(x = long, y = lat, group = group),
               color = "black", fill = "lightblue",
               size = .5, alpha = .3)
```

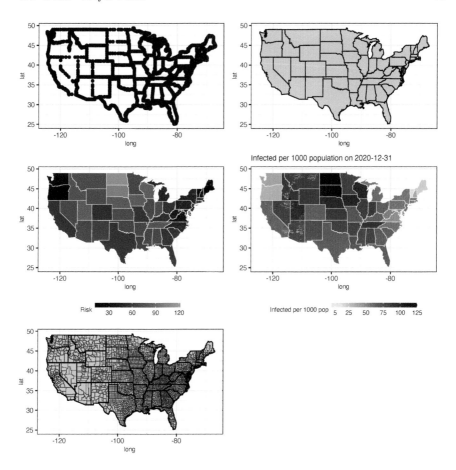

FIGURE 3.10: Various US maps. Top left: basic map with dotted state boundaries. Top right: with colored state areas. Middle left: with colored state areas according to infected per thousand population. Middle right: with colored state areas and limits on the values. Bottom left: with colored state areas and county boundaries.

3.9 Other Useful Plots

3.9.1 Density and Conditional Density Plots

Here, we consider the ILI example, and select some variables, including week_start, age_0_4, age_5_24, age_25_49, age_50_64, age_65, and to-

tal_patients between the years 2010 and 2018. Then, we change this data
format from wide to long.

```
library(cdcfluview)

Ili.usa <- ilinet(region = "national", years = NULL)

Ili.usa10 <- Ili.usa %>%
  filter(year >= 2010, year <= 2018) %>%
  select(week_start, age_0_4, age_5_24, age_25_49,
         age_50_64, age_65, total_patients)

all.df = tidyr::gather(Ili.usa10, age, ILI, -c(week_start,
  ↪  total_patients))
```

A density plot using the function `geom_density()` can be used to estimate
the density of variables, assuming that the underlying density is smooth and
continuous. The left panel of Figure 3.11 creates the basic density plots of `ILI`
for each `age` group.

```
ggplot(all.df, aes(ILI, fill = age)) +
  geom_density(alpha = 0.4, na.rm = TRUE) +
  theme(legend.position = 'bottom')
```

The conditional density plot is a useful tool for examining the relationship
between a continuous and categorical variable, since it illustrates how the
latter's conditional distribution changes over the former's value. The right
panel of Figure 3.11 provides the conditional density plots of `log(ILI + 1)`
over `age` group using `position = "fill"` and `binwidth = 1`. The size of the
bin width can be controlled by changing the value of `binwidth`; for example,
we can select `binwidth = 0.1` for a smaller bin width.

```
ggplot(all.df, aes(log(ILI + 1))) +
  geom_histogram(aes(fill = age), binwidth = 1,
                 position = "fill", na.rm = TRUE) +
  theme(legend.position = 'bottom')
```

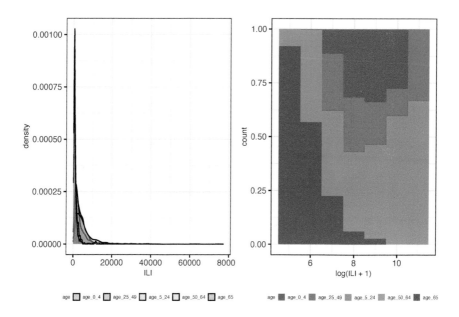

FIGURE 3.11: Left: a density plot of `ILI` for each age group. Right: a conditional density plot of `log(ILI + 1)` over `age` for `binwidth = 1`.

3.9.2 Adding Marginal Plots

As well as a scatter plot, the marginal plots would give very useful insight to make the connection between two variables and study their distributions. We can add some subplots in the x- and y-axis of a scatter plot. Before we begin with adding marginal plots, let's get ready by installing the "ggExtra" package by the following method.

```
# Install ggExtra R package
install.packages('ggExtra')
```

Let's make a simple scatter plot of `log(total_patients + 1)` versus `log(ILI + 1))` using the ILI data with a long format in the previous section. We can use the function `ggMarginal()` to add the various types of marginal plots, such as density (`type = "density"`), histogram (`type = "histogram"`), and boxplot (`type = "boxplot"`), to a original scatter plot. An example of a scatterplot with marginal histograms is given in Figure 3.12.

```
p <- ggplot(all.df, aes(log(total_patients + 1), log(ILI + 1))) +
  aes_string(color = 'age') +
  geom_point()

library(ggExtra)
ggMarginal(p, type = "histogram", color = "#FF0000", fill = "#FFA500")
```

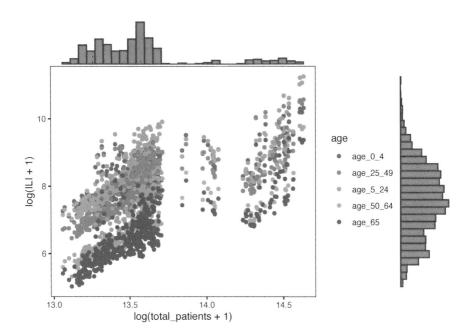

FIGURE 3.12: A scatterplot with marginal boxplots.

3.10 Arranging Plots

3.10.1 Facets

Sometimes, we wish to look that the scatterplot within each factor of categorical variables. For example, we may want to look at the situation within each Region in our case. We can split a single plot into many related plots using the function facet_wrap() or facet_grid():

- `facet_wrap(~variable)`: returns a symmetrical matrix of plots for the number of levels of variable;

- `facet_grid(. ~variable)`: returns facets equal to the levels of variable distributed horizontally.

- `facet_grid(variable~.)`: returns facets equal to the levels of variable distributed vertically.

The exemplary figures using the functions `facet_wrap()` and `facet_grid()` are illustrated in Figure 3.13.

```r
df <- IDDA::state.long %>%
  dplyr::filter(DATE == as.Date('2020-12-31'))
p <- ggplot(df, aes(log(Infected + 1), log(Death + 1))) +
  geom_point(na.rm = TRUE) +
  aes(color = Region) +
  theme(legend.position = 'bottom')

p
p + facet_grid(.~Region)
p + facet_grid(Region~.)
p + facet_wrap(~Region)
```

3.10.2 Combining Plots Using R Package "patchwork"

Before plots can be laid out, they have to be assembled. The goal of "patchwork" is to make it simple to combine separate ggplots into the same graphic. We can install patchwork from CRAN using the following.

```r
if (!require('patchwork')) install.packages('patchwork')
library(patchwork)
```

Let us consider some simple examples as below, and their corresponding plots can be found in Figure 3.14.

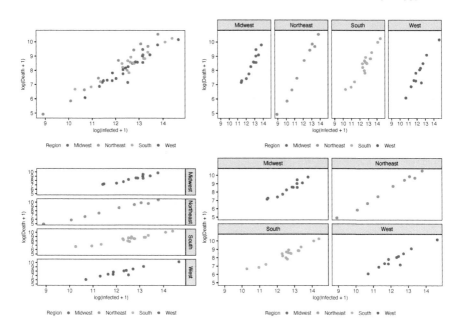

FIGURE 3.13: Faceting examples using the functions `facet_wrap()` and `facet_grid()`.

```
df   <- IDDA::state.long %>%
  dplyr::filter(DATE == as.Date('2020-12-31'))
p1 <- ggplot(df, aes(log(Infected + 1), log(Death + 1))) +
  geom_point(na.rm = TRUE)

p2 <- ggplot(df, aes(log(Death + 1))) +
  geom_histogram(binwidth = 1) + aes(fill = Region) +
  theme(legend.position = 'bottom')

p3 <- ggplot(df, aes(log(Infected + 1))) +
  geom_histogram(binwidth = 1) + aes(fill = Region) +
  theme(legend.position = 'bottom')

# Horizontal arrangement
p1 + p2
# Vertical arrangement
p1 / p2
# Grouped arrangements
p1 | (p2 / p3)
```

```
# # Combine three plots
# p1 + p2 + p3

# Set the number of plots per row
p1 + p2 + p3 + plot_layout(ncol = 2)

# Combine the duplicate legends
p1 + p2 + p3 + plot_layout(ncol = 2, guides = "collect") &
  theme(legend.position = 'bottom')
```

3.11 Saving the Figure and Output

After polishing the figure, we need to save it and output it as a readable file
for later use. We can either output it in a standard figure format, such as png,
tiff, jpeg; or we can save it as an R readable data file, usually referred to as
XXX.rds, XXX.rda or XXX.RData, and read by `readRDS('XXX.rds')`.

3.11.1 Saving in Figure Format

The function `ggsave()` is useful for saving plots. It saves the last plot presented
using the current graphics device's size by default.

```
# Take a look at the figure before saving
# print(p)
ggsave('example_ggplot2.png', p) # Save the figure in png format
```

```
## Saving 6.5 x 4.5 in image
```

3.11.2 Saving in RDS Format

The function `saveRDS()` can save a single R object to a given file (in RDS file
format), and the object can be restored using the function `readRDS()`.

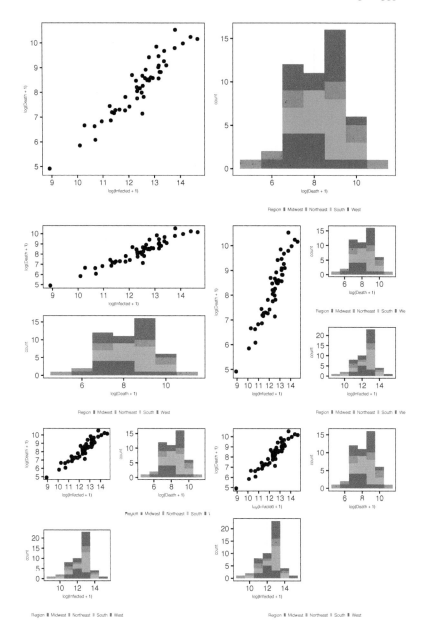

FIGURE 3.14: Patchwork examples. Top: horizontal arrangement of two plots (p1+p2). Middle left: vertical arrangement of two plots (p1/p2). Middle right: grouped arrangements of three plots (p1|(p2/p3)). Bottom left: grouped arrangements of three plots (p1+p2+p3) by setting the number of plots per row. Bottom right: grouped arrangements of three plots (p1+p2+p3), whose duplicate legends are combined through guides = "collect".

```
# Save the figure in RDS format
saveRDS(p, 'example_ggplot2.rds')
# Read the figure in RDS format
q <- readRDS('example_ggplot2.rds')
# print(q)
```

3.12 Further Reading

The web provides many valuable resources with regard to the topics covered in this chapter.

- https://ggplot2.tidyverse.org/;
- https://ggplot2-book.org/introduction.html;
- https://r-coder.com/r-graphs/;
- https://r4ds.had.co.nz/data-visualisation.html;
- https://www.r-graph-gallery.com/ggplot2-package.html;
- https://ggplot2-book.org/;
- https://datacarpentry.org/R-ecology-lesson/04-visualization-ggplot2.html
- https://patchwork.data-imaginist.com/;
- https://cran.r-project.org/web/packages/patchwork.

The major resources for the ggplot2 package can be found in the following books:

- Wickham, H. (2016). *ggplot2: Elegant graphics for data analysis.* Springer.
- Chang, W. (2018). *R graphics cookbook: Practical recipes for visualizing data.* O'Reilly Media.
- Lander, J.P. (2014). *R for everyone: Advanced analytics and graphics.* Pearson Education.

3.13 Exercises

1. Create a scatterplot using IDDA::state.long on 2020-11-01.

    (a) Create a base scatterplot. Treat Infected/1000 as the x-axis, and Death/1000 as the y-axis.

    (b) Color the points according to Division. Hint: use aes(color=).

(c) Change the size of the points to be proportional to the population. Hint: use aes(size=).

(d) Change the label of the x-axis to "Infected (in thousands)," the label of the y-axis to "Death (in thousands)."

(e) Change the title of the figure to "Infected against death on 2020-11-01."

(f) Save the plot to file "q1.png."

2. Time series plot using IDDA::state.ts for Florida.

(a) Obtain the daily new death count for Florida.

(b) Create a line plot, with time as the x-axis, and daily new deaths as the y-axis. Add the title "Daily new death count for Florida" to the plot.

(c) Using the data until November 27, 2020, a model obtained the following prediction and 80% prediction intervals from November 28 to December 11, 2020. Add another line on your time series plot indicating the predicted daily new death data. Change the title to "two-week-ahead forecast of the daily new death count for Florida" focusing your plot.

```
     DATE    Y.Death    PI
1  2020-11-28    72   [33, 111]
2  2020-11-29    56   [17,  96]
3  2020-11-30    74   [34, 114]
4  2020-12-01    88   [48, 128]
5  2020-12-02    91   [50, 131]
6  2020-12-03    59   [18, 101]
7  2020-12-04   104   [62, 146]
8  2020-12-05    79   [31, 128]
9  2020-12-06    64   [14, 113]
10 2020-12-07    81   [31, 132]
11 2020-12-08    95   [43, 148]
12 2020-12-09    98   [44, 152]
13 2020-12-10    67   [12, 122]
14 2020-12-11   111   [54, 168]
```

(d) Add ribbons on your time series plot in part (c) to illustrate the prediction intervals in part (c). Change the title to "two-week-ahead forecast of the daily new death count for Florida with 80% prediction intervals."

(e) Save the plot to file "q2.png."

3. For the data `IDDA::state.long` and focusing on November 1, 2020, do the following:

   (a) Create a map using Death per 1000 population as the coloring feature.
   (b) Save the plot to file "q3.png."

4. Combine the three figures and save the plot to file "q4.png." Hint: In R, save each plot with different names (e.g., `p1`, `p2`, `p3`), and then use the "patchwork" package.

5. Use the CDC ILI data `Ili.usa.age` in the following. Note that some age groups may be used differently through the years. For example, `age_25_49` and `age_50_64` have been used since October 4, 2009, while `age_25_64` was used until September 27, 2009.

   (a) There are missing values (N/A) and 0 `total_patients` for some rows in `all.df`. Exclude rows with 0 `total_patients` and N/A in `ILI` within `filter()`. Generate a new variable `unweighted_ILI_age` = `ILI` / `total_patients`. Make point, jittering, boxplot, and violin plots of `unweighted_ILI_age` for age groups.

   (b) Draw the density plot of `unweighted_ILI_age` for each age group.
   (c) Generate the conditional density plot of `log(ILI + 1)` grouped by age with `bandwidth` 0.5.
   (d) Make the scatterplot of `log(tota_patients + 1)` versus `unweighted_ILI_age`.
   (e) For part (d), add marginal density plots for `log(tota_patients + 1)` versus `unweighted_ILI_age`.

4

Interactive Visualization

The "plotly" is an R package to create a "plotly" object for interactivity, enabling users to engage with data in ways that are not possible with static plots. Along with interactivity, users can hover the cursor over the point or series to see specific information behind them and zoom into the regions of interest. In addition, one can auto-refresh graphics to show the most recent data in an interactive visualization. Those functionalities of the interactive plot allow users to identify patterns and trends more quickly among many comparisons and effectively communicate with the public for discussions or decisions in many research fields. For users familiar with graphic tools from the "ggplot2" package, the "plotly" package also provides the `ggplotly()` function to quickly add interactivity to the existing ggplot2 workflow by translating a "ggplot2" object to a "plotly" object.

This chapter focuses on creating and customizing the interactivity of plots based on the "plotly" package and covers various types of interactive graphics such as bar charts, scatter plots, multiple time series plots, and animation plots.

For the interactive version of the graphs presented in this chapter, see our book webpage[1].

4.1 An Introduction

Highly disparate data must be formulated to inform public health professionals and decision-makers to facilitate communication with the public and inform decisions regarding measures to protect the health of the public. Interactive data visualization allows users the freedom to explore data fully. Here are some key advantages of using interactive data visualization software:

- hovering over any data points to see the underlying data;
- identifying causes and trends more quickly;

---

[1]`https://first-data-lab.github.io/IDDA_book/interactive-visualization.html`

- adding highlights and viewing subsets of data by making use of graph options;
- auto-refreshing the visual effect to show the most recent data.

So far, our primary tool for creating these data visualizations has been "gg-plots." In the past few years, interactive tools for visualization of disease outbreaks have been improving markedly. In this chapter, we will introduce the R "plotly" package, which grants us the ability to design, share, and customize sophisticated and interactive graphics.

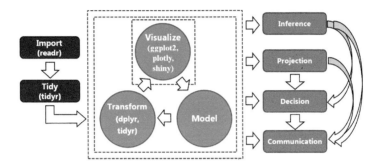

FIGURE 4.1: A typical data science process.

"Plotly" is a popular R package to produce interactive and high-quality figures, including scientific charts, maps, 3D charts, and animations. It is built on top of visualization library D3.js, HTML, and CSS. Here are some benefits of using plotly.

- Plotly is compatible with R, Python, MATLAB, Julia, etc.
- Plotly is compatible with another widely used R package "ggplot2" in R and Python.
- Plotly GUI helps users without a programming background to create interactive plots.
- Plotly produces interactive plots that can be embedded in websites using frames or HTML.
- Plotly employs intuitive syntax.

Before we begin, please get ready by installing the "plotly" R package by any of the following methods.

Installing "plotly" from CRAN

The package can be installed from R by using the written code below:

```
install.packages("plotly")
```

Installing "plotly" from GitHub

Alternatively, we can install the latest development version of "plotly" from GitHub using the R package "devtools" as follows:

```
devtools::install_github("ropensci/plotly")
```

4.2 Creating Plotly Objects

To create a plotly object, we start with a call to `plot_ly()` and pass the data. Next, we decide which graphical representation we would like to use: points, lines, bar charts, etc. Then, we can customize labels, colors, titles, fonts, etc.

Here is a typical code structure:

```
plot_ly(data) %>%
add_* (x, y, type, mode, color, size) %>%
layout(title, xaxis = list(title, titlefont),
       yaxis = list(title, titlefont))
```

In the above code, `layout()` is used to either add and/or modify part(s) of the layout of the graph. There is a family of `add_*()` functions, such as `add_histogram()`, `add_trace()`, `add_lines()`, `add_pie()`, so that we can define how to render data in geometric objects. A layer can be considered as a group of graphical elements that can be effectively summarized using the following five components: data, aesthetic mappings (how variables in the data are mapped to visual properties), a geometric representation (e.g., rectangles and circles), statistical transformations (e.g., `stat_bin()` and `stat_smooth`), and positional adjustments (e.g., `position_dodge`).

Here are some arguments that are typically used in the `add_*()` function:

- x: values for x-axis;
- y: values for y-axis;

- `type`: specify the plot that we want to create, for example, "histogram," "bar," and "scatter";
- `mode`: format in which we want the data to be represented in the plot, and possible values are "markers," "lines," "points";
- `color`: values of the same length as x, y and z representing the color of data points or lines in the plot;
- `size`: values for the same length as x, y and z representing the size of data points or lines in the plot.

4.2.1 Using `plot_ly()` to Create a Plotly Object

Before trying this example, we need to make sure to install the "plotly," "dplyr" and "lubridate" packages. The "lubridate" is an R package of choice for working with variables that store dates' values. See Appendix C for a detailed description of how to use this package.

```
library(lubridate); library(dplyr); library(plotly)
```

The county-level dataset is used to create the bar chart below. We can download the `county.top10` dataset from the IDDA package. This data contains the top 10 counties with the largest number of infected cases on December 31, 2020.

```
# install.packages("devtools")
devtools::install_github('FIRST-Data-Lab/IDDA')
```

```
library(IDDA)
data(county.top10)
county.top10
```

```
##           ID      County      State Infection Death
## 176     6037  LosAngeles California    501635  8199
## 577    17031        Cook   Illinois    346004  7282
## 334    12086  Miami-Dade    Florida    253403  3959
## 75      4013     Maricopa    Arizona    245671  4299
## 2586   48201      Harris      Texas    204850  3128
## 2542   48113      Dallas      Texas    156225  1751
```

```
## 1715 32003         Clark      Nevada    137100  1962
## 193    6071 SanBernardino California   120186  1209
## 297  12011        Broward     Florida   118512  1728
## 2705 48439        Tarrant       Texas   116931  1158
```

Now let us use plot_ly() to initialize a plotly object.

```
plot_ly(data = county.top10) %>%
  add_trace(y = ~Infection, x = ~County, type = 'bar',
            name = 'Infection')
```

After running the code, we can see a modebar in the top right-hand side of the plotly graph as in the top panel of Figure 4.2. There are several buttons appearing in the modebar. Here are a few things to try in the interactive plots:

- hovering the mouse over the plot to view the associated attributes;
- selecting a particular region on the plot and using the mouse to zoom;
- resetting the axis;
- zooming in and out.

Next, we can use layout() to modify the layout of a plotly visualization and specify more complex plot arrangements; see the bottom panel of Figure 4.2.

```
plot_ly(data = county.top10) %>%
  add_trace(y = ~Infection, x = ~County, type = 'bar',
            name = 'Infection') %>%
  layout(xaxis = list(title = "County"),
         yaxis = list(title ="Infected Count"),
         title = "Total Infected Cases on 2020-12-31")
```

We can also add text labels and annotations to a plotly project in R using add_text() given by Figure 4.3.

```
plot_ly(data = county.top10) %>%
  add_bars(y = ~Infection, x = ~County, name = 'Infection') %>%
  add_text(
    text = ~scales::comma(Infection), y = ~Infection, x = ~County,
    textposition = "top middle", showlegend = FALSE,
    cliponaxis = FALSE
  ) %>%
```

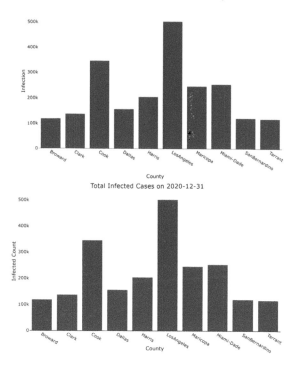

FIGURE 4.2: Top: a bar chart of the infected count. Bottom: a modified bar graph of the infected count. See the book webpage for the interactive version.

```
add_bars(y = ~Death, x = ~County, name = 'Death',
         color = I("red")) %>%
  add_text(
    text = ~Death, y = ~Death, x = ~County,
    textposition = "top middle",  showlegend = FALSE,
    cliponaxis = FALSE
  ) %>%
layout(xaxis = list(title = "County"),
       yaxis = list(title = "Number of Cases"),
       title = "Total Infected/Death Cases on 2020-12-31")
```

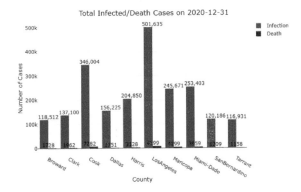

FIGURE 4.3: A bar graph of the infected count and death count.

4.2.2 Using "dplyr" Verbs to Modify Data

To visualize the states in which the counties had the most infected cases, we can use the "dplyr" verbs to modify data and calculate counts, and use add_bars() to add a new bar chart (the top panel of Figure 4.4).

```
county.top10 %>%
  group_by(State) %>%
  summarise(n = n()) %>%
  plot_ly() %>%
  add_bars(x = ~State, y = ~n)
```

Next, suppose we are interested in the distribution of the logarithm of the daily new infected cases from 2020-12-01 to 2020-12-31 from all the states in the US. We can use the state.long data in the IDDA R package, and plot the histogram of log(daily new infected cases) using add_histogram() (the bottom panel of Figure 4.4).

```
# Prepare the daily new infected count for each state
# in the period from 2020-12-01 to 2020-12-31
IDDA::state.long %>%
dplyr::filter(DATE <= '2020-12-31' & DATE > '2020-11-30') %>%
group_by(State) %>% # Group by State
# Create daily new from cum. Infected count
mutate(Y.Infected = c(Infected[-length(Infected)] -
                      Infected[-1], 0)) %>%
```

```
plot_ly() %>%
add_histogram(x = ~log(Y.Infected + 1))
```

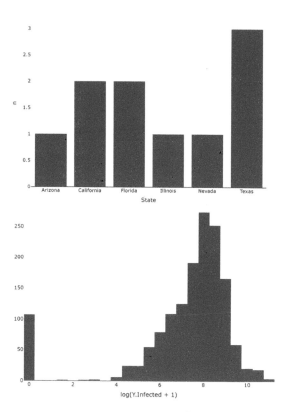

FIGURE 4.4: Top: a bar graph of the infected count by adding bars. Bottom: a histogram of the log(daily new infected cases).

4.2.3   Using `ggplotly()` to Create a Plotly Object

The ggplotly() function from the "plotly" package has the ability to translate "ggplot2" objects to "plotly" objects, which makes it really convenient to add interactivity to existing ggplot2 workflows.

We consider the state.long dataset, which includes the variables cumulative infected cases (Infected) and the cumulative number of deaths (Death). Chapter 3 shows how to draw a simple scatterplot using the reported data on December 31, 2020. The top panel of Figure 4.5 shows a translated scatterplot from "ggplot2" to "plotly."

```
df <- IDDA::state.long %>% dplyr::filter(DATE == '2020-12-31')

p <- ggplot(df, aes(log(Infected), log(Death))) +
            geom_point() +
   geom_point(aes(color = Region))
# Translate ggplot2 to plotly
ggplotly(p)
```

4.3 Scatterplots and Line Plots

The `plot_ly()` function initiates an object where one or multiple traces can be added to it via the functions `add_trace()` or `add_*()`. In `add_trace()`, the layer's type can be specified using the `type` argument. For example, some most commonly used types include `'scatter'`, `'bar'`, `'box'`, `'histogram'`, `'heatmap'`, etc. Some `add_*()` functions are specific cases of a trace type. If the type is not specified when adding a layer, a sensible default will be set.

We focus on `type = 'scatter'`, which works well in displaying lines and points, such as the time series of infected cases or the number of people vaccinated during the pandemic.

4.3.1 Making a Scatterplot

We use the `state.long` data to draw a basic scatterplot with `log(Death)` vs `log(Infected)`, as given by the middle panel of Figure 4.5.

```
library(IDDA)
data(state.long)

plot_ly(data = state.long %>%
          filter(DATE == as.Date('2020-12-31'))) %>%
  add_trace(x = ~log(Infected), y = ~log(Death), text = ~State,
            type = 'scatter', mode = 'markers')
```

4.3.2 Markers

We now describe how to change the point colors and shapes of markers generated using plotly.

- color: values mapped to relevant fill-color attribute(s);
    - I(): avoid mapping a data value to colors and specify the color manually (e.g., color = I("red")).
    - variable:
        * numeric: generate one trace with a filled color determined by the variable value and a color bar as a guide;
        * factor: generate multiple traces with different colors, one for each factor level;

- symbol: can be specified similarly as color
    - by factor value;
    - I() to set a fixed color.

- size: for scatterplots, unless otherwise specified via the sizemode, the size argument controls the minimum and maximum size of markers and must be a numeric variable.

In the bottom panel of Figure 4.5, we customize the scatterplot and change the size and color of the markers.

```
data(state.long)
plot_ly(data = state.long %>%
          filter(DATE == as.Date('2020-12-31'))) %>%
  add_trace(x = ~log(Infected), y = ~log(Death), text = ~State,
            type = 'scatter', mode = 'markers',
            # Change the size and color of the markers
            size = ~pop, color = ~Region,
            marker = list(opacity = 0.5, symbol = 'circle',
                          sizemode = 'diameter'))
```

4.3.3 A Single Time Series Plot

We draw a time series of the cumulative infected count for Cook County, IL.

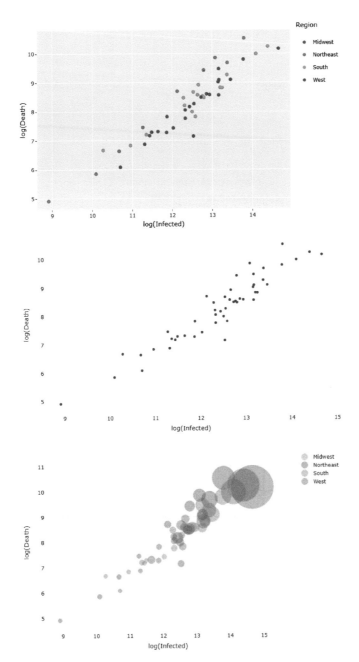

FIGURE 4.5: Top: a translated scatterplot from ggplot2 to plotly. Middle: a scatterplot with log(Death) vs log(Infected). Bottom: a customized scatterplot by changing the size and color of the markers.

```
# Load the data
library(IDDA)
data(county.top10.long)



# Start plotly from here
plot_ly() %>%
  # Add Cook County's time series using mode: lines+markers
  add_trace(data = county.top10.long %>%
    filter(wday(Date) == 1 & type == 'Observed' & County == 'Cook'),
    x = ~Date, y = ~Count, type = 'scatter', mode = 'lines+markers',
    showlegend = TRUE, name = 'mode:lines+markers')
```

4.3.4   Hovering Text and Template

We can add summary statistics or additional information to the existing plot in the form of tooltips that appear when viewers hover their mouse over areas of the plot. There are two main approaches to control the tooltip: hoverinfo and hovertemplate. The default value of hoverinfo is x+y+text+name, meaning that plotly.js will use the relevant values of x, y, text, and name to populate the tooltip text.

```
# Start plotly from here
plot_ly() %>%
  # Add Cook County's time series using mode: lines+markers
  add_trace(data = county.top10.long %>%
    filter(wday(Date) == 1 & type == 'Observed' & County == 'Cook'),
    x = ~Date, y = ~Count, type = 'scatter', mode = 'lines+markers',
    showlegend = TRUE, name = 'mode:lines+markers',
    text = 'Cook, Illinois', hoverinfo = "x+y+text")
```

To customize the tooltip on the plot, we can use hovertemplate, a template string used to render the information that appears on the hover box. See Chapter 25 of Sievert (2020) for more details on how to design and control the tooltips.

```
# Prepare hover text and formatting
label.template <-  paste('County, State: %{text}<br>',
                         'Date: %{x}<br>',
                         'Infected Cases: %{y}')



# Start plotly from here
plot_ly() %>%
  # Add Cook County's time series using mode: lines+markers
  add_trace(data = county.top10.long %>%
    filter(wday(Date) == 1 & type == 'Observed' & County == 'Cook'),
    x = ~Date, y = ~Count, type = 'scatter', mode = 'lines+markers',
    showlegend = TRUE, name = 'mode:lines+markers',
    text = 'Cook, Illinois', hovertemplate = label.template)
```

Figure 4.6 shows the time series plot of the cumulative infected count for Cook County, IL, and its customized plots using hoverinfo and hovertemplate.

4.3.5 Multiple Time Series Plots

4.3.5.1 Using Different Options in the `mode` Argument

The top panel of Figure 4.7 shows different types of time series plots for the cumulative infected count for three counties by changing the mode argument.

```
# Start plotly from here
plot_ly() %>%
  # Add Cook County's time series using mode: lines+markers
  add_trace(data = county.top10.long %>%
    filter(wday(Date) == 1 & type == 'Observed' & County == 'Cook'),
    x = ~Date, y = ~Count, type = 'scatter', mode = 'lines+markers',
    showlegend = TRUE, name = 'mode:lines+markers',
    text = 'Cook, Illinois', hovertemplate = label.template) %>%
  # Add Los Angeles County's time series using mode: lines
  add_trace(data = county.top10.long %>%
    filter(wday(Date) == 1 &
           type == 'Observed' & County == 'LosAngeles'),
    x = ~Date, y = ~Count, type = 'scatter', mode = 'lines',
    showlegend = TRUE, name = 'mode:lines',
```

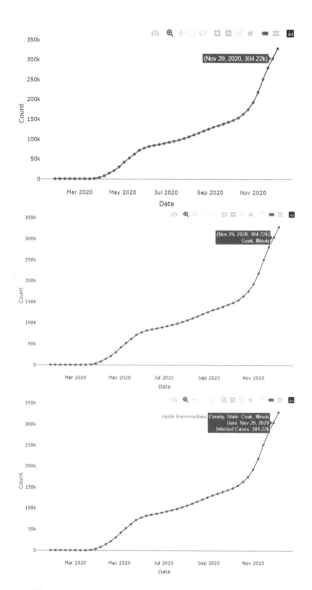

FIGURE 4.6: Top: a time series plot of the cumulative infected count for Cook County, IL. Middle: a time series plot using `hoverinfo`. Bottom: a time series plot using `hovertemplate`.

```
      text = 'Los Angeles, California',
      hovertemplate = label.template) %>%
  # Add Miami-Dada County's time series using mode: markers
  add_trace(data = county.top10.long %>%
      filter(wday(Date) == 1 &
              type == 'Observed' & County == 'Miami-Dade'),
      x = ~Date, y = ~Count, type = 'scatter', mode = 'markers',
      showlegend = TRUE, name = 'mode:markers',
      text = 'Miami-Dade, Florida', hovertemplate = label.template)
```

4.3.5.2 Mapping the Value of a Variable to Color

We can use the `color` argument to map the value of a variable to color. In the following code, we map the color of time series to different counties; see the middle panel of Figure 4.7.

```
plot_ly() %>%
  add_trace(data = county.top10.long %>%
      filter(wday(Date) == 1 & type == 'Observed'),
      x = ~Date, y = ~Count, type = 'scatter',
      mode = 'lines+markers',
      color = ~County,
      showlegend = TRUE)
```

4.3.5.3 Controlling the Color Scale

We can use the `colors` argument to control the color scale:

- "colorbrewer2.org" palette name (e.g., "YlOrRd" or "Blues");
- a color interpolation function such as `colorRamp()`;
- a vector of colors to interpolate in hexadecimal with the code format "#RRGGBB".

For example, we can define our own color palette:

```
mycol <- c("#5B1A18", "#F21A00", "#D67236", "#F1BB7B", "#D8B70A",
           "#A2A475", "#81A88D", "#78B7C5", "#3B9AB2", "#7294D4",
           "#C6CDF7", "#E6A0C4")
```

In the following code, we change the color of time series to our own defined color palette; see the bottom panel of Figure 4.7.

```
plot_ly() %>%
  add_trace(data = county.top10.long %>%
              filter(wday(Date) == 1 & type == 'Observed'),
            x = ~Date, y = ~Count, type = 'scatter',
            mode = 'lines+markers', color = ~factor(County),
            colors = mycol, showlegend = TRUE)
```

4.3.6 More Features About the Lines

We can also alter the thickness of the lines in the time series plot, and make them dashed or dotted using default types or self-defined methods. In the following code, we change the line type by the value of variable `type` by `linetype = ~type`; see the top panel of Figure 4.8.

```
plot_ly() %>%
  add_trace(data = county.top10.long %>%
              filter(County == 'Cook'),
            x = ~Date, y = ~Count, type = 'scatter',
            mode = 'lines', linetype = ~type,
            showlegend = TRUE, text = 'Cook, Illinois',
            hovertemplate = label.template)
```

4.3.7 Adding Ribbons

We can use the `add_ribbons()` function to draw a filled area plot, for example, the confidence band or prediction intervals. Its main arguments are:

- `data`: the data;
- `x`: x values;
- `ymin`: the lower bound of the ribbon;
- `ymax`: the upper bound of the ribbon.

The following code adds the 80% prediction intervals for the cumulative infected cases for Cook County, Illinois; see the bottom panel of Figure 4.8.

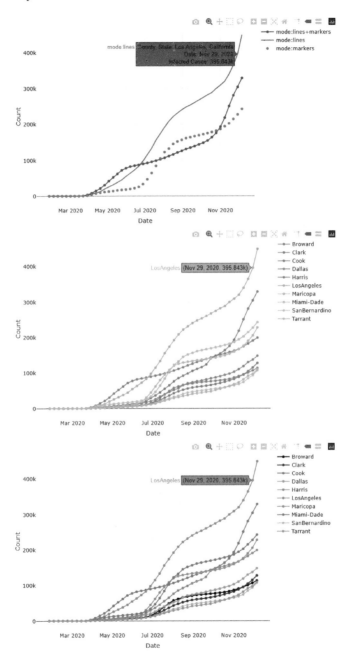

FIGURE 4.7: Top: a time series plot of the cumulative infected count for three counties using different options in the mode argument. Middle: a time series plot of the cumulative infected count by mapping the value of a variable to color. Bottom: a time series plot of the cumulative infected count by controlling the color scale.

```
plot_ly(data = county.top10.long %>%
            filter(County == 'Cook')) %>%
    add_trace(x = ~Date, y = ~Count, type = 'scatter',
             mode = 'lines', linetype = ~type,
             showlegend = TRUE, text = 'Cook, Illinois',
             hovertemplate = label.template) %>%
    add_ribbons(x = ~Date, ymin = ~Count_lb, ymax = ~Count_ub,
                 color = I("#74A089"), opacity = 0.75,
                 name = "80% prediction intervals")
```

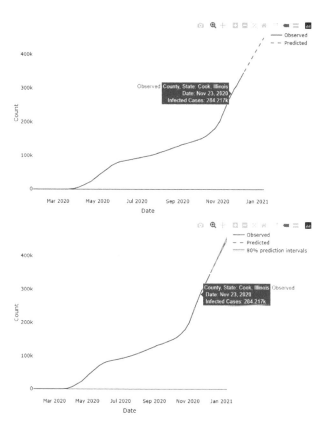

FIGURE 4.8: Top: a time series plot of the cumulative infected count and predictions for Cook County, IL, with more features about the lines. Bottom: an example of adding the ribbons to the previous time series plot.

4.4 Pie Charts

We then demonstrate how to make static and interactive pie charts in R. To draw the pie chart, we download the `features.state` from the IDDA R package, and the dataset contains four variables: State, Region, Division and pop. We are interested in the composition of the population in each region.

4.4.1 Draw Static Pie Charts

In the following, we will try the "ggplot2" package to draw the pie chart. Before starting to draw the plot, we need to prepare the data first.

```
# Prepare the data
features.region <- features.state %>%
  group_by(Region) %>%
  summarize(tpop = sum(pop))
df <- features.region %>%
  arrange(desc(Region)) %>%
  mutate(prop = round(tpop / sum(features.region$tpop), 4) * 100) %>%
  mutate(lab.pos = cumsum(prop)- 0.5 * prop )
```

Next, we will apply the `geom_bar()` and `coord_polar()` functions together with `ggplot()` to display the pie chart; see Figure 4.9.

```
# Draw the pie chart using ggplot
mycols <- c("#0073C2FF", "#EFC000FF", "#868686FF", "#CD534CFF")
ggplot(df, aes(x = "", y = prop, fill = Region)) +
  geom_bar(width = 1, stat = "identity", color = "white") +
  coord_polar("y", start = 0)+
  geom_text(aes(y = lab.pos,
                label = paste(prop, "%", sep = "")), color = "white")+
  scale_fill_manual(values = mycols) +
  theme_void()
```

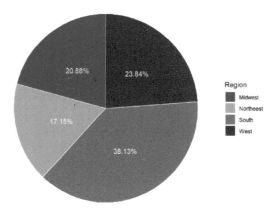

FIGURE 4.9: A simple ggplot pie chart for population in different regions.

4.4.2   Drawing Interactive Pie Charts

Now, we will try to use `plot_ly` to make the pie chart. Note that the function `add_pie()` can be implemented easily without data preparation; see Figure 4.10.

```
library(IDDA)
data(features.state)
fig1 <- plot_ly(features.state) %>%
  add_pie(labels = ~Region, values = ~pop)
fig1
```

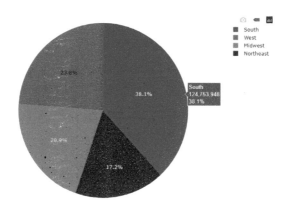

FIGURE 4.10: An interactive pie chart for population in different regions.

Next, we are interested in finding the composition of the cumulative infected/death cases in each region using `add_pie()`. We can create pie chart subplots by using the domain attribute. Here the x and y arrays set the horizontal and vertical positions, respectively. For example, x=[0, 0.5], y=[0, 0.5] indicate the bottom left position of the plot; see Figure 4.11.

```r
fig2 <- plot_ly(data = state.long %>%
  filter(DATE == as.Date('2020-12-31')) )

fig2 <- fig2 %>%
  # Add the first pie chart
  add_pie(labels = ~Region, values = ~Infected,
          name = "Region",
          domain = list(x = c(0, 0.4), y = c(0.4, 1))) %>%
  # Add the second pie chart
  add_pie(labels = ~Region, values = ~Death,
          name = "Region",
          domain = list(x = c(0.6, 1), y = c(0.4, 1))) %>%
  # Change the title and other features of the layout
  layout(title = "Pie Charts with Subplots: Infected/Death Count",
         showlegend = F,
         xaxis = list(showgrid = F, zeroline = F,
                      showticklabels = F),
         yaxis = list(showgrid = F, zeroline = F,
                      showticklabels = F))
fig2
```

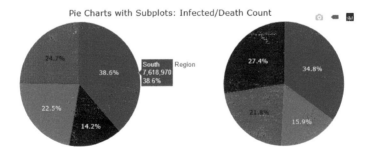

FIGURE 4.11: Pie charts with subplots. Left: infected count. Right: death count.

4.5   Animation

Animated plots are a great way to display the dynamics of the underlying data. Both `plot_ly()` and `ggplotly()` can produce key frame animations through the `frame` argument/aesthetic. They also support an `ids` argument to guarantee smooth transitions between objects with the same id during the animation. This section provides a walk-through for creating an animated plot using the "plotly" R package.

4.5.1   An Animation of the Evolution of Infected vs. Death Count

Figure 4.12 creates an animation of the evolution in the relationship between the state-level logarithm of the cumulative infected count and the logarithm of the cumulative death count evolved over time. For simple illustration purposes, we only show the evolution in December of 2020. The data `state.long` from the IDDA package provides a daily time series for 48 mainland states and the District of Columbia in the US. Below, we first prepare the data:

```
#install_github('FIRST-Data-Lab/IDDA')
library(IDDA)
data(state.long)
state.long.DEC <- state.long %>%
  dplyr::filter(DATE > as.Date("2020-11-30")) %>%
                mutate(log.Infected = log(Infected + 1)) %>%
                mutate(log.Death = log(Death + 1))
```

Next, we load the required packages.

```
# Load the required packages:
library(ggplot2); library(plotly)
```

We can use the `frame` argument in `plot_ly()` or the frame ggplot2 aesthetic in `ggplotly()` to create animations. Animated plots can be generated with the `frame =` and `ids =` arguments in the `geom_point()` function when using `ggplot()` and `ggplotly()`. In this case, the data `state.long` is recorded on a daily basis, so we will assign the DATE variable to `frame`; each point in the scatterplot represents a state, so we will assign the State variable to ids, which ensures a smooth transition from date to date for the lower 48 states and the District of Columbia in the US:

```
gg <- ggplot(state.long.DEC,
            aes(log.Infected, log.Death, color = Region,
                frame = as.factor(DATE), ids = State)) +
  geom_point(aes(size = pop))
anim1 <- ggplotly(gg)
anim1
```

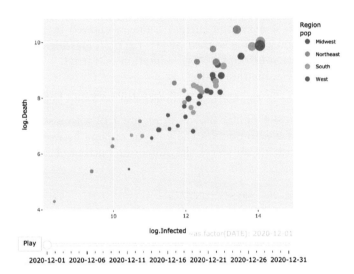

FIGURE 4.12: Our first animated plot between logarithms of the death count and infected count. See the book webpage for the animation.

Given that frame = is provided, an animation will be produced with play/pause button(s) as well as a slider component to control the animation. By default, animations populate a play button and slider component that can be used to control the state of the animation. We can pause an animation by clicking on a relevant location on the slider bar. We can remove or customize these components using the animation_button() and animation_slider() functions.

Following the rules specified by animation_opts(), we can control the play button and slider component transition between frames. Moreover, the animation_opts() function offers a few other animation options, such as the amount of time between frames, the smooth transition duration, and the type of transition easing. Here are some animation configuration options in the function animation_opts():

- p: a plotly object.
- `frame`: the duration in milliseconds of each frame (this amount should include the `transition`).
- `transition`: the duration of the smooth transition between frames, in milliseconds. If `transition = 0`, then the updates will be synchronous.
- `easing`: the type of transition easing, for example, `linear`, `quad`, `cubic`, `elastic`, `sin` and others.
- `redraw = TRUE`: trigger a redraw of the plot at the completion of the transition or not. This is desirable for transitions that include properties that cannot be transitioned but may significantly slow down updates that do not require a full redraw of the plot.
- `mode`: describes how a new animate call interacts with currently running animations: if `immediate`, current animations are interrupted and new animation is started; if `next`, the current frame is allowed to complete, after which the new animation is started; if `afterall`, all existing frames are animated to completion before the new animation is started.
- `hide`: removes the animation slider.

The detailed options in the above arguments can be found here[2].

Figure 4.13 illustrates a similar plot as in Figure 4.12 with a slightly different aesthetic style, and it also doubles the amount of time between frames, uses elastic transition easing, and places the animation buttons closer to the slider.

```
base <- state.long.DEC %>%
  plot_ly(x = ~log.Infected, y = ~log.Death, size = ~pop,
          text = ~State, hoverinfo = "x+y+text") %>%
  layout(xaxis = list(type = "log"))

anim2 <- base %>%
  add_markers(color = ~Region, frame = ~DATE, alpha = 0.8,
              span = I(2), ids = ~State, colors = "Set1") %>%
  animation_opts(frame = 1000, easing = "elastic", redraw = FALSE) %>%
  animation_button(
    x = 1, xanchor = "right", y = 0, yanchor = "bottom"
  ) %>%
  animation_slider(
    currentvalue = list(type = "date", font = list(color="red"))
  )
anim2
```

In the above code, the `span` argument describes the width of the stroke with default zero, and `span = I(2)` sets the width to be around two pixels. The

---

[2]`https://github.com/plotly/plotly.js/blob/master/src/plots/animation_attributes.js`

options x = 1, xanchor = "right", y = 0 and yanchor = "bottom" indicate that the right (bottom) side of the animation plot will be positioned at x = 1 (y = 0). The speed at which the animation progresses is controlled by the `frame` argument, with the default value being 500 milliseconds. In this example, we increase it to 1000 milliseconds, resulting in slower transitions between frames. We can change the way of transition from frame to frame via the `easing` argument. Here `easing` = "elastic" causes the points to bounce when a new frame occurs. The `redraw` = FALSE argument can improve laggy animation performance by not entirely redrawing the plot at each transition. However, in this example, it doesn't make much difference.

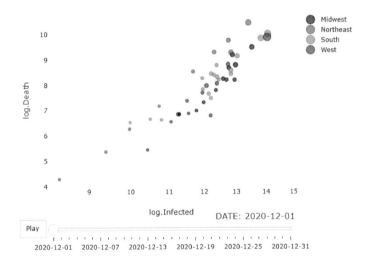

FIGURE 4.13: A modified animation with `frame` = 1000 and elastic easing. See the book webpage for the animation.

4.5.2 An Animation of the State-level Time Series Plot of Infected Count

We now would like to show the state-level time series plot of the infected count. We then show the animation by `Region`, and because there is no meaningful relationship between objects in different frames of Figure 4.14, the smooth transition duration is set to 0. This eliminates any ambiguity about whether the smooth transitions are connected in any way. Note that these options affect all animations triggered by the play button and those triggered by the slider.

```
mycol <- c("#5B1A18", "#F21A00", "#D67236", "#F1BB7B", "#D8B70A",
           "#A2A475", "#81A88D", "#78B7C5", "#3B9AB2", "#7294D4",
           "#C6CDF7", "#E6A0C4")

base <- state.long %>%
  mutate(log.Infected = log(Infected + 1)) %>%
  plot_ly(x = ~DATE, y = ~log.Infected, frame = ~Region,
          text = ~State, hoverinfo = "text") %>%
  add_lines(color = ~factor(State),
            colors = mycol, showlegend = FALSE)

anim3 <- base %>%
  layout(xaxis = list(type = "date",
                      range=c('2020-01-22', '2020-12-31'))) %>%
  animation_opts(1000, easing = "elastic",
                 redraw = FALSE, transition = 0)

anim3
```

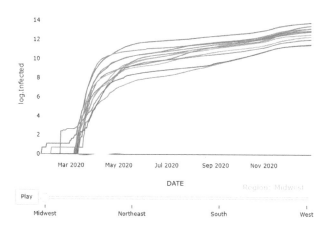

FIGURE 4.14: Animation of a time series plot of infected count by region.
See the book webpage for the animation.

4.6   Saving HTMLs

After polishing the figure, we need to save the figure and animation for later
use. We can save any widget made from any "htmlwidgets" package (e.g.,
plotly, leaflet, etc.) as a standalone HTML file using the saveWidget() function.
It generates a completely self-contained HTML file by default, with all of the
necessary JavaScript and CSS dependency files bundled inside.

4.6.1   Saving as Standalone HTML Files

The following example shows how to save an object to a standalone HTML
file.

```
# Save plotly object into a standalone html file
library(htmlwidgets)
saveWidget(fig1, "pie1.html", selfcontained = T)
saveWidget(anim1, "anim1.html", selfcontained = T)
```

4.6.2   Saving as Non-self-contained HTML Files

Sometimes, we may want to embed multiple widgets in a larger HTML document and save all the dependency files in a single directory externally. We
can do this by setting selfcontained = FALSE and specifying a fixed libdir in
saveWidget().

```
# Save plotly object into a non-selfcontained html file
library(htmlwidgets)
saveWidget(fig2, "pie2.html", selfcontained = F, libdir = "lib")
saveWidget(anim2, "anim2.html", selfcontained = F, libdir = "lib")
```

4.7   Further Reading

A good introduction to the use of plotly and plotly examples can be found on
the web.

- `https://plotly.com/r/;`
- `https://cran.r-project.org/web/packages/plotly/;`
- `https://plotly-r.com/.`

For a good reference book for the use of plotly, see:

- Sievert, C. (2020). *Interactive web-based data visualization with R, plotly, and shiny.* CRC Press. Available at `https://plotly-r.com.`

4.8   Exercises

1. We will explore the basic functions of `plot_ly` using `state.long`. Install the GitHub R package IDDA.

   ```
   library(IDDA)
   data(state.long)
   ```

   (a) Create a bar graph for the top ten states with the largest number of new infected cases on December 31, 2020.
   (b) Create a time series plot for the logarithm of the cumulative infected cases for Florida.
   (c) Create a time series plot for the logarithm of the cumulative infected cases for the top ten states with the largest number of new infected cases on December 31, 2020.
   (d) Create a pie chart for the daily new infected cases on December 31, 2020, for different regions.
   (e) Save the above plots as an HTML file, and save all the dependency files externally into a single directory using `htmlwidgets::saveWidget()` with `selfcontained = FALSE`.

2. Redraw the above plots in Parts a, b and c for the death count using the `ggplotly()` function. Save each of the above plots as an HTML file, and save all the dependency files externally into a single directory using `htmlwidgets::saveWidget()` with `selfcontained = FALSE`.

3. During the COVID-19 pandemic, we are interested in how many tests are coming back positive. The `state.long` dataset comprises state-level cumulative tests. We want to create an animation to demonstrate the weekly test positivity rate for each state based on a 7-day moving average. It is calculated by dividing the state's new

positive counts in the past seven days by the state's new tests in the past seven days. The PosTest.state.rda and Test.state.rda are the daily reported positive test and daily total test data collected from the COVIDTracking Project (https://covidtracking.com/), and they can be downloaded from the IDDA R package.

(a) Load the datasets to your working directory:

```
library(IDDA)
data(Test.state)
data(PosTest.state)
```

Change them from the wide form to the long form, and combine them into one dataset.

(b) Add a new column for the weekly test positivity rate.
(c) Create an animated time series plot for Florida's weekly test positivity rate in the past month (30 days) starting from December 2 to December 31, 2020. For example, if we pause on December 7, 2020, it should show a time series of Florida's weekly test positivity rate from November 8 to December 7, 2020.
(d) Save your animation in part c as a standalone HTML file and a non-self-contained HTML file.

5

R Shiny

Thanks to interactive visualization, users can understand and use complex data in infectious disease data learning. Shiny allows us to create a graphical user interface (GUI) that can be used locally or remotely. It has great potential for making interactive, web-based visualizations much more accessible to users. For example, it provides multiple views or panels, allowing users to examine their data from various perspectives. Shiny is also useful for displaying and communicating updated findings to a large audience.

This chapter will look at how to connect "plotly" graphs to "shiny," an open-source R package that provides an elegant and powerful web framework for creating R-based web applications. Shiny allows us to turn the data into interactive web applications without having to know HTML, CSS, or JavaScript.

For the interactive version of the Shiny applications presented in this chapter, see our book webpage[1].

5.1 An Introduction to Shiny

Shiny is available on CRAN, so we can install it in the usual way from the R console:

```
install.packages("shiny")
```

5.1.1 Structure of a Shiny Application

The classic structure of Shiny applications is illustrated in Figure 5.1. To create a Shiny application, we start with a directory that contains two files: **ui.R** and **server.R**.

---

[1]https://first-data-lab.github.io/IDDA_book/shiny.html

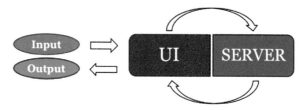

FIGURE 5.1: An illustration of Shiny structure.

- In **ui.R**, we define the user interface, UI, which determines how input and output widgets appear on a page. It takes input from users, passes it to the server, and displays the output that is returned from the server. The UI can be highly customized. Packages like "shinydashboard" make it simple to build interfaces with advanced layouts. Historically, we used function `shinyUI()` to register a UI with Shiny. Now we only need to ensure that the last expression in **ui.R** is a user interface.

- In **server.R**, we define the server function, which establishes a link between input and output widgets. Specifically, it takes input from the UI, generates output, and returns output that can be displayed by the UI. Historically, we used the function `shinyServer()` for the server function to be recognized by Shiny. Now we only need to ensure that the last expression in **server.R** is an appropriate server function. In several examples of this chapter, we still use `shinyUI()` and `shinyServer()` to better illustrate the structure of Shiny applications.

We can also save both server and UI functions in a single file called **app.R** in the directory without separating them into **server.R** and **UI.R**.

When necessary, additional data, scripts, and other resources can be put under the same directory to support the application.

5.1.2 Launching a Shiny Application

We can execute a Shiny application in two ways:

- We can run `runApp(<path to directory>)` in the R console. Here, `<path to directory>` is the path of the directory that contains **ui.R** and **server.R** mentioned above. We will illustrate it in detail in the examples later in this chapter. Similarly, if we started with a directory with only one file, **app.R**, we can run `runApp(<path to directory>)`, where `<path to directory>` is the path of the directory that contains **app.R**. See the structure of the file as follows.

app.R

```
library(shiny)
# Define the UI
ui <- shinyUI(pageWithSidebar(
      # ... Program here
  ))

# Define the server logic
server <- shinyServer(function(input, output) {
      # ... Program here
})
# Run the application
shinyApp(ui = ui , server = server)
```

- If we decide to run a Shiny application using commands in the RStudio console, we can simply call `shinyApp(ui, server)` after defining the server function as `server()`, and the UI function as `ui()`.

In practice, it can be beneficial to use two separate files **ui.R** and **server.R** to manage complicated `ui()` and `server()` functions. Otherwise, saving the code in one file **app.R** is convenient.

5.1.3 Creating the First Shiny Application

Now let's write our first Shiny application that showcases the histogram of simulated normal random variables, where we can adjust the number of simulations using the UI. This example is borrowed from and hosted on Shiny[2]. Readers are referred to the website to interact with the Shiny application.

The user interface is defined in the **ui.R** source file. The main component is the `sliderInput()` function, which enables the user to pass an input to Shiny by sliding a bar in the UI.

ui.R

```
library(shiny)
# Define the UI control
shinyUI(pageWithSidebar(
  # Application title
```

---
[2]https://shiny.rstudio.com/gallery/example-01-hello.html

```
  headerPanel("Hello Shiny!"),
  # Sidebar with a slider input for the number of observations
  sidebarPanel(
    # Set the variable name of input as "obs"
    sliderInput("obs",
                # Set the greeting sentence on the sliderInput
                "Number of observations:",
                # Set the range of the sliderInput
                min = 1,
                max = 1000,
                # Set the default selection of the sliderInput
                value = 500)
  ),
  mainPanel(
    # Show output, plot of generated distribution, in the mainPanel
    plotOutput("distPlot")
  )
))
```

The application's server function is defined in the file **server.R** below. In
the server function, we retrieve the value of input through sliderInput() by
calling input$obs. Then a random distribution with the desired number of
observations is generated, and the histogram is plotted. Then we use function
renderPlot() to wrap the histogram plot so it can be returned to the UI and
displayed.

server.R

```
# Define the server logic
shinyServer(function(input, output) {
  # Expression that generates a plot of the distribution
  # The expression is wrapped in a call to renderPlot()
  # to indicate that:
  #  (i)  It is "reactive" and therefore should be automatically
  #       re-executed when inputs change
  #  (ii) Its output type is a plot
  output$distPlot <- renderPlot({
    # Generate normal random variables and plot the histogram
    dist <- rnorm(input$obs)
    hist(dist)
  })
})
```

If everything is working properly, executing the Shiny application in RStudio would result in a popup html window as in Figure 5.2.

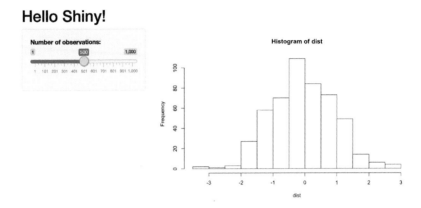

FIGURE 5.2: The hello Shiny example. This example is hosted on https://shiny.rstudio.com/gallery/example-01-hello.html.

5.1.4 Creating a New Shiny Application in RStudio

In RStudio, we can create a new directory and an **app.R** file containing a basic app in one step by clicking File -> New File -> Shiny Web App, then providing an "Application Name" and selecting an "Application Type." See the RStudio interface in Figure 5.3.

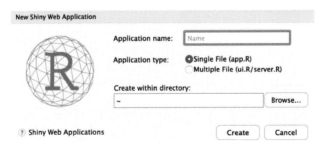

FIGURE 5.3: The RStudio interface for creating a new Shiny web application.

5.1.5 Sharing the Shiny Application

After creating the Shiny applications, we can share them for others to view and use. There are two options: (i) sharing Shiny applications as R scripts, and (ii) hosting a webpage that embeds Shiny applications.

The first option, sharing as R scripts, is great for developers who also use R because it enables them to reuse the code. To do this, we can either send a zip folder that contains all files needed or host the zip folder online and send the link to the zip folder. If we choose to do the latter, other users can simply run `runUrl("<the weblink>")` to view the Shiny application. In addition, we can also create a repository on GitHub to host the files, and other users can run `runGitHub("<repository name>", "<username>")` in the console of RStudio to view the Shiny applications.

In this book, we share our Shiny application examples through R scripts on the "RShinyapp" sub-directory of the GitHub repository "IDDA_book" of user "FIRST-Data-Lab". Please run the following to retrieve each example.

```
runGitHub(repo = "IDDA_book", username = "FIRST-Data-Lab",
          subdir = "RShinyapp/<Shiny application name>", ref = "main")
```

The second option, sharing as a webpage, is suitable for sharing with the general public due to its accessibility. RStudio offers three ways to host the Shiny application as a webpage: (i) create a free or professional account at `http://shinyapps.io`, a cloud-based service from RStudio, to host the Shiny applications; (ii) use RStudio Connect by clicking the Publish icon in the RStudio IDE ($>=0.99$) or run `rsconnect::deployApp("<path to directory>")`; (iii) use Shiny Server, a companion program to Shiny that builds a web server designed to host Shiny applications. It's free, open-source, and available from GitHub.

5.2 Useful Input Widgets

In addition to `slideInput()` in the previous example, Shiny supports a number of other useful input widgets (or web elements) with which the user can interact. Specifically, when the user selects or inputs values through the input widgets, the input will be passed to the Shiny application. Then the output would be generated and shown in the user interface instantly.

Below is a list of basic input widgets with the interface shown in Figure 5.4.

- `selectInput()` or `selectizeInput()` for dropdown menus;
- `numericInput()` for a single number;
- `sliderInput()` for a numeric range;
- `textInput()` for a character string;
- `dateInput()` for a single date;
- `dateRangeInput()` for a range of dates;
- `fileInput()` for uploading files;
- `checkboxInput()`, or `checkboxGroupInput()` or `radioButtons()` for choosing a list of options.

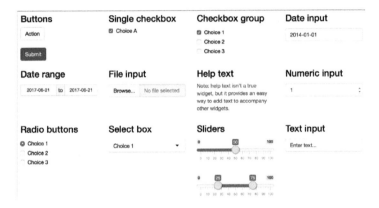

FIGURE 5.4: The basic input widgets. See the widgets section in https://shiny.rstudio.com/gallery/ for more examples.

Input widgets can be easily stylized with CSS and/or SASS, and even customized input widgets can be integrated, despite the fact that many Shiny applications use them "out of the box." From `https://shiny.rstudio.com/`, we can explore and select the appropriate input widgets for the interactive visualization.

In the following, we will concentrate on using "plotly" and static R graphics as inputs to other output widgets, rather than using these input widgets to link multiple graphs in Shiny through direct manipulation.

5.3 Displaying Reactive Outputs

We can create reactive outputs with the following two-step process.

Step 1: Add an R object to the UI.

Shiny offers a family of functions to turn R objects into output for the UI. Each function produces a specific type of output. An output function creates

- `dataTableOutput`: DataTable;
- `htmlOutput`: raw HTML;
- `imageOutput`: image;
- `plotOutput`: plot;
- `tableOutput`: table;
- `textOutput`: text;
- `uiOutput`: raw HTML;
- `verbatimTextOutput`: text.

We can add output to the user interface in the same way we add HTML elements and widgets. The output function can be put inside `sidebarPanel()` or `mainPanel()` in the `ui()`.

Step 2: Provide R code to build the object.

Next, we notify Shiny how to build the object by providing the R code that builds the object in the server function. In the Shiny process, the server function creates a list-like object called output that contains all of the code needed to update the R objects in the app. In the list, each R object must have its own entry. We can make an entry by defining a new output element in the server function, as shown below. The name of the element should be the same as the name of the reactive element we created in the `ui()`. We do not need to explicitly state in the server function's last line of code that it should return output. R uses reference class semantics to update output automatically.

The output of one of Shiny's `render*()` functions should be included in each entry to output. These functions take an R expression and perform some basic pre-processing on it. The `render*()` function that corresponds to the type of reactive object we are creating is used. Specifically, the render function generates the following:

- `renderDataTable()`: DataTable;
- `renderImage()`: images (saved as a link to a source file);
- `renderPlot()`: plots;
- `renderPrint()`: any printed output;
- `renderTable()`: data frame, matrix, other table like structures;
- `renderText()`: character strings;
- `renderUI()`: a Shiny tag object or HTML.

Each `render*()` function takes a single argument: an R expression surrounded by {}, and it can be as simple as a single line of text or as complex as a function call with many lines of code.

This R expression can be thought of as a set of instructions that we give Shiny to remember. When the application is first launched, Shiny will run the instructions with default input, and then it will run them again every time the input changes.

The Shiny Cheatsheet[3] provides a nice summary of the render*() and *Output() functions; see Figure 5.5.

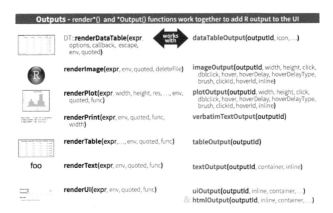

FIGURE 5.5: The basic output widgets. Source: https://shiny.rstudio.com/ images/shiny-cheatsheet.pdf.

5.4 Rendering Plotly inside Shiny

The renderPlotly() function renders anything that the plotly_build() function understands, for example, plot_ly(), ggplotly(), and "ggplot2" objects. It also renders NULL as an empty HTML <div>, which can be convenient sometimes when rendering a graph doesn't make sense.

Example 5.1. (Top Ten States with the Highest Cumulative Infected Counts). In this example, we develop a Shiny application to show a barplot of the top 10 states with the highest cumulative infected COVID-19 counts on a given day selected using the slider bar.

ui.R

---

[3] https://shiny.rstudio.com/images/shiny-cheatsheet.pdf

```
library(plotly); library(shiny)
library(dplyr); library(IDDA)

fluidPage(
  sliderInput("date.update",
              label = h5("Select date"),
              min = as.Date("2020-12-01"),
              max = as.Date("2020-12-31"),
              value = as.Date("2020-12-31"),
              timeFormat = "%d %b",
              animate = animationOptions(interval = 2000, loop = FALSE)
  ),
  # Show a plot of the generated distribution
  mainPanel(
    plotlyOutput("state_cum_inf", height = "100%", width = "150%")
  )
)
```

server.R

```
state.cum.inf <- function(date.update){
  d.str = as.character(format(date.update, 'X%Y.%m.%d'))
  # Order cumulative infected counts for each state
  dat = IDDA::I.state %>% arrange(desc(get(d.str)))
  # Select the top 10 states with highest counts
  dat.top <- dat[1:10, ] %>%
    dplyr::select(State, all_of(d.str)) %>%
    mutate(Date = date.update)
  names(dat.top) <- c('State', 'CumInf', 'Date')
  plot.title <- paste0("Cumulative infected counts on ",
                       as.character(date.update))
  # Set up the order of bars
  dat.top$State = factor(dat.top$State,
                  levels = dat.top$State[order(dat.top$CumInf,
                                               decreasing = T)])
  # Draw the bar plot
  plot.out <-
    ggplot(dat.top, aes(x = State, y = CumInf)) +
    labs(title = plot.title) +
    geom_bar(stat = 'identity', fill = "#C93312") +
    xlab('') +
    ylab('')
```

```
   return(plot.out)
}

function(input, output) {
   output$state_cum_inf <- renderPlotly({
      ts <- state.cum.inf(input$date.update)
   })
}
```

Starting from this example, we omit `shinyUI()` and `shinyServer()` to align
with the latest Shiny. If everything is working correctly, we will see the application appear in the browser, looking something like the top panel of Figure
5.6. For the result of the R Shiny application in this example, one can run the
R command below:

```
runGitHub(repo = "IDDA_book", username = "FIRST-Data-Lab",
          subdir = "RShinyapp/state_cum_inf", ref = "main")
```

Example 5.2. Time Series Plot of the Top Ten Counties. This example
demonstrates the time series plot of the risk of the top ten counties based on
the corresponding measurement on December 31, 2020. For the time series
plot, the user can choose different risk measurements: the infection rate (IR),
standardized infection ratio (SIR), and the weekly local risk (WLR). The IR
is the number of cases and the population divided by the population in the
county. The SIR standardizes the data by re-expressing them as the ratio
between the observed number of cases and the number that would have been
expected in a standard population. It is the infected cases per population
divided by country infected cases per population. The WLR is another risk
measure, which is the average number of new cases in the past week per 100,000
people in a county. The WLR gives the user a way to assess a county's current
risk level compared to other counties.

ui.R

```
fluidPage(
    div(class="outer",
    tags$head(includeCSS("styles.css")),
    plotlyOutput("county_risk_ts", height="100%", width="100%"),
    absolutePanel(id = "control", class = "panel panel-default",
       top = 60, left = 70, width = 255, fixed=TRUE,
```

```
      draggable = TRUE, height = "auto", style = "opacity: 0.8",
      selectInput("plot_type",
        label = h5("Select type"),
        choices = c("WLR" = "wlr", "IR" = "localrisk",
                    "SIR" = "sir")
      ) # End of selectInput
    ) # End of absolutePanel
  ) # End of div
)
```

server.R

```
date.update <- as.Date('2020-12-31')
mycol <- c("#5B1A18", "#F21A00", "#D67236", "#F1BB7B",
           "#D8B70A", "#A2A475", "#81A88D", "#78B7C5",
           "#3B9AB2", "#7294D4", "#C6CDF7", "#E6A0C4")
ts.plotly = function(df, type = 'scatter', mode = 'lines+markers',
                     group = group, mycol, showlegend = TRUE, visible = T,
                     xaxis = xaxis, yaxis = yaxis, legend = legend) {

  ts <- plot_ly(df) %>%
    add_trace(x = ~x, y = ~y, type = type, mode = mode,
              color = ~group, colors = mycol,
              showlegend = showlegend,
              visible = visible) %>%
    layout(xaxis = xaxis, yaxis = yaxis,
           legend = legend)
  return(ts)
}

county.risk.ts = function(date.update, type = 'localrisk'){
  date.all = date.update - (0:29)
  date.lag = date.all - 7
  County.pop0 <- IDDA::pop.county
  County.pop <- County.pop0 %>%
    filter((!(State %in% c("Alaska","Hawaii"))))
  County.pop <- County.pop %>%
    filter((!(ID %in% c(36005, 36047, 36081, 36085))))
  County.pop$ID[County.pop$ID == 46102] = 46113
  dat <- IDDA::I.county
  dat <- dat %>%
    filter((!(State %in% c("Alaska", "Hawaii"))))
```

```
var.names <- paste0("X", as.character(date.all), sep = "")
var.names <- gsub("\\-", "\\.", var.names)
var.lag <- paste0("X", as.character(date.lag), sep = "")
var.lag <- gsub("\\-", "\\.", var.lag)
tmp <- as.matrix((dat[, var.names] - dat[, var.lag])/7)
dat <- dat[, c("ID", "County", "State", var.names)]
sir.c <- sum(County.pop$population)/as.matrix(colSums(dat[,
↪   -(1:3)])))

I0 <- LogI0 <- LocRisk0 <- SIR0 <- dat
LogI0[,-(1:3)] <- as.matrix(log(dat[,-(1:3)]+1))
# Infection rate per thousand population
LocRisk0[,-(1:3)] <- sweep(as.matrix(dat[, -(1:3)]), 1,
                           County.pop$population[match(dat$ID,
↪   County.pop$ID)],
                           "/") * 1000
# Standardized infection rate (%)
SIR0[,-(1:3)] <- sweep(LocRisk0[, -(1:3)], 2, sir.c/1000*100, "*")
# Weekly local risk
WLR0 <- sweep(tmp, 1,
              County.pop$population[match(dat$ID, County.pop$ID)],
              "/") * 1e5

county.dat <- data.frame(Date = date.all)
CountyState <- paste(as.character(dat$County),
                     as.character(dat$State), sep = ",")
LocRisk <- cbind(county.dat, round(t(LocRisk0[, -(1:3)]), 2))
names(LocRisk) <- c("Date", CountyState)
SIR <- cbind(county.dat,t(SIR0[, -(1:3)]))
names(SIR) <- c("Date", CountyState)
WLR <- cbind(county.dat, round(t(WLR0), 2))
names(WLR) <- c("Date", CountyState)

xaxis.fr <- list(title = "", showline = FALSE,
                 showticklabels = TRUE, showgrid = TRUE,
                 type = 'date', tickformat = '%m/%d')
legend.fr <- list(orientation = 'h', x = 0, y = -0.05,
                  autosize = F, width = 250, height = 200)

if (type == 'localrisk'){
  ind.county = order(LocRisk0[, var.names[1]], decreasing = TRUE)
  df.fr <- LocRisk %>%
    select(c(1, 1 + ind.county[1:10])) %>%
```

```
        gather(key = "County.State", value = "LogI", -Date)
      names(df.fr) <- c("x","group","y")
      yaxis.fr <- list(title = "Infection Rate (Cases per Thousand)")
      ts.fr <- ts.plotly(df.fr, type = 'scatter',
                          mode = 'lines+markers',
                          group = group, mycol,
                          showlegend = TRUE, visible = T,
                          xaxis = xaxis.fr, yaxis = yaxis.fr,
                          legend = legend.fr)
    }else if (type == 'sir'){
      ind.county = order(SIR0[, var.names[1]], decreasing = TRUE)
      df.fr <- SIR %>%
        select(c(1, 1 + ind.county[1:10])) %>%
        gather(key = "County.State", value = "LogI", -Date)
      names(df.fr) <- c("x", "group", "y")
      yaxis.fr <- list(title = "SIR (%)")
      ts.fr <- ts.plotly(df.fr, type = 'scatter',
                          mode = 'lines+markers',
                          group = group, mycol, showlegend = TRUE,
                          visible = T, xaxis = xaxis.fr,
                          yaxis = yaxis.fr, legend = legend.fr)
    }else if (type == 'wlr'){
      ind.county <- order(WLR0[, var.names[1]], decreasing = TRUE)
      df.fr <- WLR %>%
        select(c(1, 1 + ind.county[1:10])) %>%
        gather(key = "CountyState", value = "WLR", -Date)
      names(df.fr) <- c("x", "group", "y")
      yaxis.fr <- list(title = "WLR (New Cases Per 100K)")
      ts.fr <- ts.plotly(df.fr, type = 'scatter',
                          mode = 'lines+markers',
                          group = group, mycol,
                          showlegend = TRUE, visible = T,
                          xaxis = xaxis.fr, yaxis = yaxis.fr,
                          legend = legend.fr)
    }
  return(ts.fr)
}


function(input, output) {
  output$county_risk_ts <- renderPlotly({
    ts <- county.risk.ts(date.update, type = input$plot_type)
  })
}
```

If everything is working correctly, we will see the application appear in the browser, which looks like the middle panel of Figure 5.6. For the result of the R Shiny application in this example, one can run the R command below:

```
runGitHub(repo = "IDDA_book", username = "FIRST-Data-Lab",
          subdir = "RShinyapp/county_risk", ref = "main")
```

Example 5.3. ILI Data. Various Types of Plots for Age Groups.
We build a Shiny application to show various types of plots for different age groups in this example. Users can choose a scatterplot, jitter plot, boxplot, and violin plot.

ui.R

```
library(shiny); library(dplyr); library(plotly)
library(tidyr); library(lubridate); library(cdcfluview)

fluidPage(
    div(class="outer",
    plotlyOutput("ili_age_scatter", height="100%", width="100%"),

    absolutePanel(id = "control", class = "panel panel-default",
                  top = 300, left = 85, width = 255, fixed=TRUE,
                  draggable = TRUE, height = "auto",
                  style = "opacity: 0.8",
    selectInput("plot_type",
      label = h5("Select type"),
      choices = c("Scatter" = "Scatter",
                  "Jitter" = "Jitter",
                  "Box" = "Box",
                  "Violin" = "Violin")
    )
    )
    )
)
```

server.R

```
mycol <- c("#F21A00","#7294D4", "#D67236", "#78B7C5", "#D8B70A")
ili_age_scatter = function(date.update, plot.type) {
```

```
Ili.usa <- ilinet(region = "national", years = NULL)

Ili.usa10 <- Ili.usa %>%
  filter(year >= 2010, year <= 2018) %>%
  select(week_start, age_0_4, age_5_24, age_25_49, age_50_64,
         age_65, total_patients)

all.df = tidyr::gather(Ili.usa10, age, ILI,
                       -c(week_start, total_patients))

all.df$age[which(all.df$age == "age_0_4")] = "age_00_04"
all.df$age[which(all.df$age == "age_5_24")] = "age_05_24"

if (plot.type == 'Scatter'){
  g <- ggplot(all.df, aes(age, log(ILI + 1), color = age)) +
         geom_point()
} else if (plot.type == 'Jitter'){
  g <- ggplot(all.df, aes(age, log(ILI + 1), color = age)) +
         geom_jitter()
}else if (plot.type == 'Box'){
  g <- ggplot(all.df, aes(age, log(ILI + 1), color = age)) +
       geom_boxplot()
} else if (plot.type == 'Violin'){
  g <- ggplot(all.df, aes(age, log(ILI + 1), color = age)) +
       geom_violin()
}
  return(g)
}


function(input, output) {
  output$ili_age_scatter <- renderPlotly({
    g <- ili_age_scatter(date.update = date.update,
                    plot.type = input$plot_type)
  })
}
```

If everything is working correctly, we will see the application appear in the browser, looking something like the bottom panel of Figure 5.6. For the result of the R Shiny application in this example, one can run the R command below:

```
runGitHub(repo = "IDDA_book", username = "FIRST-Data-Lab",
          subdir = "RShinyapp/ili_age_scatter", ref = "main")
```

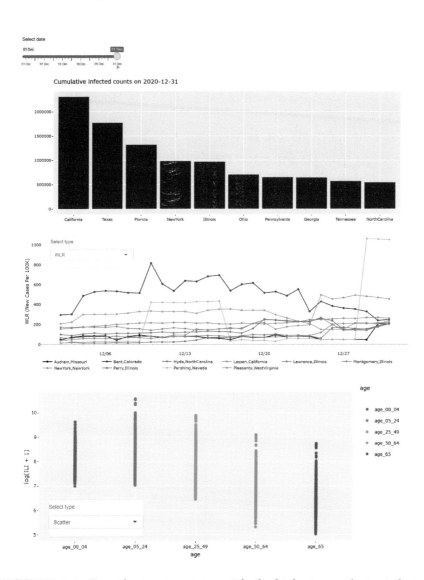

FIGURE 5.6: Top: the top ten states with the highest cumulative infected count. Middle: a time series plot of the top ten counties (based on the chosen measurement on December 31, 2020). Bottom: various types of plots (scatter, jitter, box, and violin) for different age groups in the ILI data.

5.5   Further Reading

The webpage of R Shiny provides a wealth of information:

- https://shiny.rstudio.com/.

Shiny Gallery provides a myriad of Shiny apps to be inspired by and learn from.

- https://shiny.rstudio.com/gallery/.

The following webinars give a very good introduction to Shiny:

- https://shiny.rstudio.com/tutorial/;

- https://www.rstudio.com/resources/webinars/introduction-to-shiny/.

For further information on the Shiny R package introduced in this chapter, see:

- Wickham, H. (2021). *Mastering Shiny: Build interactive apps, reports, and dashboards powered by R.* O'Reilly Media, Inc.

- Sievert, C. (2020). *Interactive web-based data visualization with R, plotly, and shiny.* CRC Press.

6

Interactive Geospatial Visualization

In the last two decades, many spatial and spatiotemporal methods for early outbreak detection, cluster detection, risk area and factor identification, and disease transmission pattern evaluation have been developed, boosting the study of spatial epidemiology. By definition, the focus of spatial epidemiology is the study of the geographical or spatial distribution of health outcomes. It is sometimes interchangeably known as disease mapping. Usually, it has the incidence of disease or prevalence of disease as its main focus. It is commonplace to consider a geographic dimension included within a research design in infectious disease studies, which may involve initial visualization of the distribution and some simple summary measures. The growing development of the open-source community has aided the application of spatial epidemiology methods, with R, a programming language and free software for statistical computing and graphics, being the most widely used and popular.

To display data on a variety of COVID-19 metrics, interactive geospatial visualization is used by websites ranging from the *New York Times* and the *Washington Post* to GitHub and Flickr. This chapter explores the interactive geospatial visualization of the data. It introduces the "leaflet" R package and illustrates basic usages and properties with several examples of disease mapping at state and county levels. The chapter closes with an illustration of integrating "leaflet" with R Shiny.

For the interactive and colored version of the graphs and applications presented in this chapter, see our book webpage[1].

6.1   An Introduction to Leaflet

One of the most popular open-source JavaScript libraries for interactive maps is "leaflet." GIS specialists such as "OpenStreetMap," "Mapbox," and "CartoDB" use it frequently. This package makes integrating and controlling map objects in R simple.

---

[1]https://first-data-lab.github.io/IDDA_book/spatial-visualization.html

6.1.1 Features and Installation

Unlike static visualization R packages like "ggplot2" or "ggmap," the maps constructed by the R package "leaflet" are fully interactive and can include features like interactive panning, zooming, popups, tooltips and labels, as well as highlighting and selecting regions.

6.1.1.1 Features

The R package "leaflet" uses the JavaScript library "Leaflet" and the "html-widgets" package to create and customize interactive maps. It can be used to compose maps using arbitrary combinations of map tile, markers, polygons, lines, popups and GeoJSON.

We can use "leaflet" to render spatial objects generated from R packages "sp" and "sf," as well as data frames with latitude and longitude columns with ease. The maps created by "leaflet" can be interactively viewed directly from the R console, RStudio, Shiny applications, and R Markdown documents. They can be bonded together with mouse events to drive "Shiny" logic and thus help develop interactive R Shiny applications. Plugins from the "leaflet" plugins repository are useful to enhance map features.

6.1.1.2 Installation

To install the R package "leaflet," run the following command at the R prompt:

```
# Install from R CRAN
install.packages("leaflet")
# Install the development version from GitHub
devtools::install_github("rstudio/leaflet")
```

6.1.2 Basic Usage

Similar to "ggplot2," leaflet maps are built using layers. We can create a Leaflet map with these basic steps:

Step 1. Create a base map widget by calling leaflet().

Step 2. Customize the map widget using layer functions (addTiles(), addMarkers(), etc.) to add features to the map.

Step 3. Print the map widget to display it and save it.

Here is a base example. Suppose we would like to show the Eiffel Tower on a street map; we can run the following:

```r
library(leaflet)
m <- leaflet() %>%
  addTiles() %>%  # Add default OpenStreetMap map tiles
  addMarkers(lng = 2.2945, lat = 48.8584, popup = "The Eiffel Tower")
```

```r
m  # Print the map
```

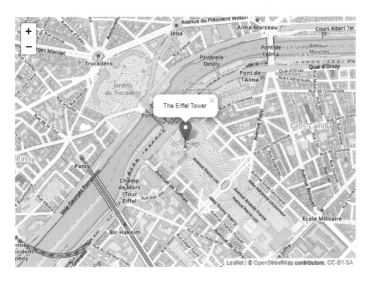

FIGURE 6.1: An example of `leaflet()` displaying the Eiffel Tower.

6.2 The Data Object

The data parameter in both the `leaflet()` and map layer functions is designed to receive spatial data in one of several formats from:

- The base R:
  - lng-lat matrix;
  - data frame with lng-lat matrix.

- The "sp" package:
  - SpatialPolygons, SpatialPolygonsDataFrame, Polygons, and Polygon objects;
  - SpatialLines, SpatialLinesDataFrame, Lines, and Line objects.
- The "map" package:
  - the data frame returned from `map()`.

The `data` argument is used to send spatial data to functions that need it; for example, if `data` is a "SpatialPolygonsDataFrame" object, then calling the `addPolygons()` function on that map widget will add the polygons from the "SpatialPolygonsDataFrame." We demonstrate how to use these three methods in the following.

6.2.1   Specifying Latitude/Longitude in Base R

By providing `lng` and `lat` arguments to the layer function, we can always explicitly identify latitude/longitude columns.

Alternatively, we can provide a data frame that consists of the latitude/longitude information. For example, in the `addCircles()` below, we directly pass a data frame, `df`, with variables `Lat` and `Long` in the data frame:

```
# Add ten circles to specified locations
df <- data.frame(Lat = rexp(10) + 40, Long = rnorm(10) - 92)
leaflet(df) %>% addTiles() %>%
  addCircles(data = df, lat = ~Lat, lng = ~Long)
```

6.2.2   Using R Package "sp"

The first general R package to provide classes and methods for spatial data is called "sp," which provides classes and methods to create points, lines, polygons, and grids and operate on them. For example, we can generate the polygon objects using the function `Polygon()`, and we can also generate "SpatialPolygons" objects using lists of `Polygon()`.

```
library(sp)
library(rgeos)
x1 <- c(3, 3, 6, 12, 3)
x2 <- c(6, 3, 2, 6, 6)
```

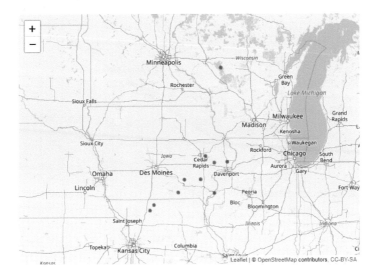

FIGURE 6.2: An example of addCircles() to include variables Lat and Long in the data frame.

```
y1 <- c(6, 3, 2, 6)
y2 <- c(2, 3, 2, 2)
Poly1 <- Polygon(cbind(x1, x2))
Poly2 <- Polygon(cbind(y1, y2))
Polys1 <- Polygons(list(Poly1), "s1")
Polys2 <- Polygons(list(Poly2), "s2")
SPolys <- SpatialPolygons(list(Polys1, Polys2), 1:2)
```

To draw this in leaflet, we can use addPolygons():

```
leaflet(height = "300px") %>% addPolygons(data = SPolys)
```

6.2.3   Using R Package "maps"

The R "maps" package contains many outlines of continents, countries, states, and counties, for example, the world database and the USA databases (usa, state, county). We can check help(package='maps') for the complete list. We can specify map(fill = TRUE) for polygons and FALSE for polylines. The code

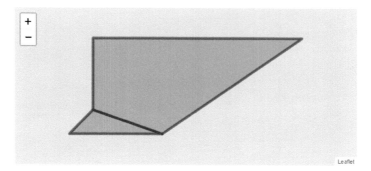

FIGURE 6.3: An example of `addPolygons()` to draw the polygons objects.

below shows how to obtain and plot the geospatial object of the states in the US; see the left panel of Figure 6.4.

```
library(maps)
mapStates <- map("state", fill = TRUE, plot = FALSE)
```

```
leaflet(data = mapStates) %>%
  addTiles() %>%
  addPolygons(fillColor = topo.colors(10),
              opacity = 0.5, stroke = FALSE)
```

6.3   Choropleth Maps

A **choropleth map** is a map that colors or patterns a set of pre-defined areas in proportion to a statistical variable that represents an aggregate summary of a geographic characteristic within each area, such as population, different numbers, or disease rates.

Let us start by loading data from JSON into "sp" objects using the "geojsonio" package to create a choropleth map. This will make manipulating geographic features and their properties in R much easier.

```
library(geojsonio)
urlRemote  <- "https://raw.githubusercontent.com/"
pathGithub <- "PublicaMundi/MappingAPI/master/data/geojson/"
fileName   <- "us-states.json"
states0 <- geojson_read(x = paste0(urlRemote, pathGithub, fileName),
                        what = "sp")
class(states0)
```

In Chapter 2, we have seen an example of merging the `states0` data with
`state.long` in the IDDA R package, and the combined data is saved as `states1`
in the IDDA package. In the following, we will work with `states1` directly. Now
let us load the required R packages and data to our working directory.

```
library(geojsonio); library(leaflet); library(dplyr)
library(IDDA); data(state.long); data(states1)
# Remove the following regions due to lack of data
states1 <- states1 %>%
  subset(!name %in% c('Alaska', 'Hawaii', 'Puerto Rico'))
```

6.3.1 Creating a Base Map

First, we will create a basic US state map. The easiest way to add tiles is by
calling `addTiles()` with no arguments:

```
dmap <- leaflet(states1) %>%
        setView(-96, 37.8, 4, zoom = 4) %>%
        addTiles()
```

Here, the `setView()` function sets the view of the map (center and zoom level)
with the following arguments:

- `lng`: the longitude of the map center;
- `lat`: the latitude of the map center;
- `zoom`: the zoom level.

Besides the `setView()` function, there are other methods to manipulate the
map widget, for example:

- `fitBounds(map, lng1, lat1, lng2, lat2)`: sets the bounds of a map;

- setMaxBounds(map, lng1, lat1, lng2, lat2): restricts the map view to the given bounds;
- clearBounds(map): clears the bounds of a map, which are automatically determined, if available, from the latitudes and longitudes of the map elements (otherwise, the full world view is used).

Next, using the function addPolygons() with no additional arguments, we can obtain the uniform polygons with default styling without any customization; see the right panel of Figure 6.4.

```
dmap %>% addPolygons()
```

FIGURE 6.4: Left: the geospatial object of the states in the US. Right: the uniform polygons with default styling without any customization.

6.3.2   Coloring the Map

Next, we design the color palette for the map. The function colorFactor() converts numeric or factor/character data values to colors using a palette that can be provided in a variety of formats. Two often-used arguments are

- palette: the colors or color function that values will be mapped to;
- domain: the possible values that can be mapped.

```
pal.state.factor <- colorFactor(
  palette = "YlOrRd", domain = states1$Division)
```

Now, let us color the states according to the division that they belong to. In this case, we can map the value of the division to colors using `fillColor = ~pal.state.factor(Division)`. We can also customize the map, change the color, line type of the state boundary, and other style properties; see the top left panel of Figure 6.5.

```
dmap %>% addPolygons(
         fillColor = ~pal.state.factor(Division),
         weight = 1, opacity = 1,
         color = "white", dashArray = "3",
         fillOpacity = 0.9, layerId = ~name_ns)
```

6.3.3 Interactive Maps

On the interactive choropleth map, it is possible to zoom in and hover over a state to get more details about it. The next step will be to highlight the polygons as the mouse passes over them. The highlight argument in the addPolygon() function makes this simple.

We will generate the highlight labels using the `sprintf()` function, a wrapper for the C library function with the same name. `sprintf()` returns a character vector with a formatted mix of text and variable values.

```
labels_cases <- sprintf(
        "<strong>%s</strong><br/>Population: %g M<br>
        Cumulative Cases: %g<br>Death: %g<br>
        Infected Cases per Thousand: %g",
        states1$name, round(states1$pop / (1e6), 2),
        states1$Infected,
        states1$Death, states1$Infect_risk * 1000) %>%
   lapply(htmltools::HTML)

labels_cases[[1]]
```

In the above `sprintf()` function, each `%` is referred to as a slot, which is basically a placeholder for a variable that will be formatted. The letter `s` indicates that the formatted variable is specified as a string. `%g` indicates that the formatted variable uses compact decimal or scientific notation.

The `labelOptions` argument of the `addPolygons()` function allows us to customize marker labels. To emphasize the currently moused-over polygon, we will use the highlightOptions() function. The labelOptions argument can be populated using the labelOptions() function. Now let us display the state names and values to the user; see the top right panel of Figure 6.5.

```
dmap <- dmap %>% addPolygons(
        fillColor = ~pal.state.factor(Division),
        weight = 1, opacity = 1,
        color = "white", dashArray = "3",
        fillOpacity = 0.9, layerId = ~name_ns,
        # Options to highlight a shape on hover
        highlight = highlightOptions(
          weight = 5, color = "#666",
          dashArray = NULL, fillOpacity = 0.9,
          bringToFront = TRUE),
        # Add labels
        label = labels_cases,
        labelOptions = labelOptions(
        style = list("font-weight" = "normal", padding = "3px 8px"),
         textsize = "15px", direction = "auto"))
```

When the mouse is hovered over a state, the label is displayed. The bringToFront = TRUE in the highlightOptions() function is used to prevent the active polygon's thicker, white border from being hidden behind the borders of other polygons. Later in this chapter, we will introduce the **z-index**, which determines the order of layers.

Finally, let us add the legend using the function addLegend() (The bottom panel of Figure 6.5).

```
dmap <- dmap %>% addLegend(pal = pal.state.factor, values = ~Division,
                opacity = 0.7, title = NULL,
                position = "bottomright")
```

6.4   Legends

6.4.1   Classification Schemes

The legend of a map can be used to list the features on the map and what they represent. Symbols in the legend should be the same as they are in the

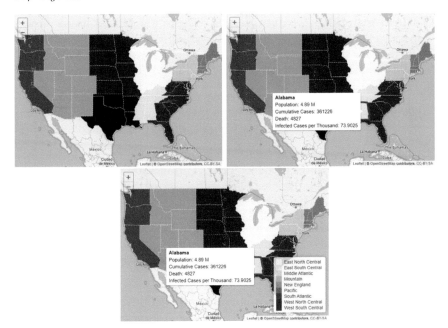

FIGURE 6.5: Top left: the customized polygons with different colors and lines of the state boundary. Top right: the moused-over polygons with state names and information. Bottom: the moused-over polygons with state names and information and the legend for the regions.

body of the map. In infectious disease studies, we often observe the following types of data:

- **Nominal**: variables are nominal if they are given names or titles to differentiate one entity from another, such as the name of a place, whether urban or rural.

- **Ordinal**: variables are ordinal if the order of their value matters but not the difference between values; for instance, the risk level of a disease may be classified with Level 1 representing the lowest risk, Level 2 second-lowest, and so on).

- **Numeric**: variables are numeric if the measurement has a numerical meaning; examples of numeric data include temperature, population density, male-to-female ratio, the number of infected cases. Numeric values may vary on a discrete (e.g., integer) or continuous scale.

According to Pfeiffer et al. (2008), the continuous attribute data can be divided into six basic classification schemes:

- **Natural groupings of data values**: Classes are defined based on what appear to be natural groupings of data values. Breakpoints that are known to be relevant to a particular application, such as fractions and multiples of mobility levels or risk thresholds, may be used to define the breaks.

- **Quantile breaks**: The data is divided into a set of classes, each with an equal number of observations. Quintile (five-category) classifications, for example, are ideal for displaying data that is linearly distributed.

- **Breaks with equal intervals**: The attribute value's range is calculated and divided into evenly spaced intervals. This method is useful for mapping data that is uniformly distributed or if the user of the map is familiar with the data ranges (e.g., herd sizes or temperature bands).

- **Standard deviation classifications**: This method uses the number of standard deviations above and below the mean to calculate the distance between the observation and the mean. It is best for data that are normally distributed.

- **Arithmetic progressions**: At an arithmetic (additive) rate, the widths of category intervals increase in size. If the first category is one unit wide and the width is increased by one unit, the second category becomes two units wide, the third three units wide, and so on (1, 3, 6, ...). This method is particularly useful when dealing with data from **skewed** distributions.

- **Geometric progressions**: The widths of the category intervals are multiplicatively increased at a geometric rate. For example, if the first category's interval width is two units, the second category will be $2 \times 2 = 4$ units wide, the third $2 \times 2 \times 2 = 8$ units wide, and so on. This method can also be used to analyze data from **skewed** distributions.

6.4.2   Mapping Variables to Colors

We demonstrate how to apply the above classification schemes to map values to colors. For simplicity, we wrap the above code into a function and run it with different palettes, data, labels, variables, etc.

```
map.state <- function(dat, fill.var, labels, pal, ID = 'name_ns'){
  dmap <- leaflet(dat) %>%
    setView(-96, 37.8, 4, zoom = 4) %>%
    addTiles() %>%
    addPolygons(
      fillColor = ~pal(dat@data %>% pull(fill.var)),
      weight = 1, opacity = 1, color = "white",
```

```
      dashArray = "3", fillOpacity = 0.9,
      layerId = ~dat@data %>% pull(ID),
      highlight = highlightOptions(
        weight = 5, color = "#666", dashArray = NULL,
        fillOpacity = 0.9, bringToFront = TRUE),
      label = labels,
      labelOptions = labelOptions(
        style = list("font-weight" = "normal", padding = "3px 8px"),
        textsize = "15px", direction = "auto")) %>%
    addLegend(pal = pal, values = ~dat@data %>%
              pull(fill.var), opacity = 0.7,
              title = NULL, position = "bottomright")
  dmap
}
```

The family of `color*()` can be used to generate palette functions easily. There are currently three color functions for dealing with continuous input: `color-Numeric()`, `colorBin()`, and `colorQuantile()`; and one for categorical input, `colorFactor()`. Each function has two required arguments:

- `palette`: specifies the colors to map the data to;
- `domain`: specifies the range of input values.

The following are some examples, applying `colorNumeric()`, `colorBin()`, and `colorQuantile()` to the `states1` data.

The top left and right panels of Figure 6.6 display the maps produced using `colorNumeric()` and `colorQuantile()`, respectively. The bottom left and right panels of Figure 6.6 illustrate the maps produced using `colorBin()` and `colorFactor()`, respectively.

The `colorNumeric()` function is a simple linear mapping from continuous numeric data to an interpolated palette.

```
pal.state.numeric <- colorNumeric(
  palette = "YlOrRd", domain = states1$Infected)

map.state(dat = states1, fill.var = 'Infected', labels = labels_cases,
          pal = pal.state.numeric, ID = 'name_ns')
```

The `colorBin()` function also maps continuous numeric data, but performs binning based on value (see the `cut()` function).

```
bins.state<- c(0, 1e4, 5e4, 1e5, 5e5, 1e6, 5e6)
pal.state.bins <- colorBin("YlOrRd", domain = states1$Infected,
                            bins = bins.state)

map.state(dat = states1, fill.var = 'Infected',
          labels = labels_cases,
          pal = pal.state.bins, ID = 'name_ns')
```

The `colorQuantile()` function similarly bins numeric data, but via the `quantile()` function.

```
pal.state.quantile <- colorQuantile(
  palette = "YlOrRd", domain = states1$Infected, n = 8)
map.state(dat = states1, fill.var = 'Infected', labels = labels_cases,
          pal = pal.state.quantile, ID = 'name_ns')
```

The `colorFactor()` function maps factors to colors. If the palette is discrete and has a different number of colors than the number of factors, interpolation is used.

```
pal.state.factor <- colorFactor("YlOrRd", domain = states1$Region)

map.state(dat = states1, fill.var = 'Region',
          labels = labels_cases,
          pal = pal.state.factor, ID = 'name_ns')
```

6.5   Examples of County-level Maps

We are interested in the infection rate and COVID-19 related control policies at the county level. In this section, we demonstrate how to draw a county-level choropleth map to illustrate the spatial variation from county to county by making use of the datasets `counties1` and `counties2` from the IDDA package.

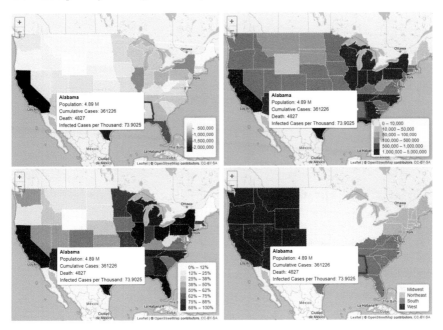

FIGURE 6.6: Examples of mapping variables to colors. Top left: `colorNumeric()`. Top right: `colorBin()`. Bottom left: `colorQuantile()`. Bottom right: `colorFactor()`.

6.5.1 A County-level Map of COVID-19 Infection Risk

Before constructing the map, we briefly demonstrate how the `counties1` dataset is obtained. First, let's download the raw spatial data from the online repository of the "plotly" package, and name it `counties0`. The raw data is in the "json" format, which is widely used for transmitting structured data in web applications.

```
urlRemote   <- "https://raw.githubusercontent.com/"
pathGithub  <- "plotly/datasets/master/"
fileName    <- "geojson-counties-fips.json"
counties0 <- geojson_read(x = paste0(urlRemote, pathGithub, fileName),
                          what = "sp")
```

Then we add the state-level information from the dataset `IDDA::states1`, the infected and death count on December 31, 2020, from `IDDA::I.county` and `IDDA::D.county`, the policy on May 1, 2020 from `IDDA::policy.county` to the raw county-level data.

```
data(I.county)
data(D.county)
counties1 <- counties0
counties1@data <- left_join(counties1@data, states1@data %>%
  select(id, name, density, name_ns, Region, Division, pop, DATE),
  by = c('STATE' = 'id'))

names(counties1)[8] <- 'state_name'
names(counties1)[10] <- 'state_name_ns'

counties1$id <- as.integer(counties1$id)

counties1@data <- left_join(counties1@data, pop.county %>%
  select(ID, population), by = c("id" = "ID"))

counties1@data <- left_join(counties1@data, I.county %>%
  select(ID, X2020.12.31), by = c("id" = "ID"))

counties1@data <- left_join(counties1@data, D.county %>%
  select(ID, X2020.12.31), by = c("id" = "ID"))
  names(counties1@data)[16:17] <- c('Infected', 'Death')
  names(counties1@data)[13] <- 'pop_state'
  counties1@data <- counties1@data %>%
  mutate(Infect_risk = Infected/population)

counties1[counties1$id == 46113, 'population'] <- 14309
```

Then we obtain the dataset `IDDA::counties1`, which can also be loaded from the IDDA package directly. Now let us load the data and put together the color palette, scheme, and the highlight label.

```
# Load the data directly
data(counties1)
data(states1)
counties1 <- counties1 %>%
  subset(!state_name %in% c('Alaska', 'Hawaii', 'Puerto Rico'))
states1 <- states1 %>%
  subset(!name %in% c('Alaska', 'Hawaii', 'Puerto Rico'))

# Prepare the color palette and scheme
col2 <- colorRampPalette(c("#053061", "#2166AC", "#4393C3",
                            "#92C5DE","#D1E5F0", "#FFFFFF",
```

```
                    "#FDDBC7", "#F4A582", "#D6604D",
                    "#B2182B", "#67001F"))

pal.county.quantile <- colorQuantile(
  palette = col2(200), domain = counties1$Infect_risk, n = 8)

# Prepare the highlight label
labels_cases.county <- sprintf(
  "<strong>%s</strong>, <strong>%s</strong>
  <br/>Infection Rate: %g <br>
  Population: %g K <br>
  Infected Cases on 2020-12-31: %g<br>
  Death Cases on 2020-12-31: %g",
  counties1$NAME, counties1$state_name,
  round(counties1$Infect_risk, 3),
  counties1$population / 1000,
  counties1$Infected, counties1$Death
) %>% lapply(htmltools::HTML)
```

Next, we draw the county-level map. In "leaflet," **map panes** group layers together implicitly, which allows web browsers to work with multiple layers at once, saving time over working with layers individually. The **z-index** CSS property is used in map panes to always show some layers on top of others. The following is the default priority order (least to greatest): tile layers and grid layers, paths (such as lines, polylines, circles, or GeoJSON layers), marker shadows, marker icons, and popups.

We can use the function addMapPane() to customize map panes for a leaflet map to control layer order, and we can specify:

- name: the name of the new pane;
- zIndex: the zIndex of the pane. Panes with a higher index are rendered first, followed by panes with a lower index.

By setting the "pane" argument in leaflet options, we can use this "pane" to render overlays (points, lines, and polygons). This will allow us to control the layer order; for example, points will always be on top of polygons, and states will always be on top of counties.

The output of the following county-level map is shown in the top panel of Figure 6.7.

```r
# Draw the county-level map
dmap2 <- leaflet() %>%
  setView(-96, 37.8, zoom = 4) %>%
  addTiles() %>%
  # Add additional panes to control layer order
  # Display borders (zIndex: 420) above the polygons (zIndex: 410)
  addMapPane("polygons", zIndex = 410) %>%
  addMapPane("borders", zIndex = 420) %>%
  # Add state polygons (borders pane)
  addPolygons(
    data = states1, fill = FALSE, weight = 1,
    color = "gray", fillOpacity = 0,
    options = pathOptions(pane = "borders")
  ) %>%
  # Add county polygons (polygons pane)
  addPolygons(
    data = counties1,
    fillColor = ~pal.county.quantile(Infect_risk),
    weight = 1, opacity = 1, color = "white",
    dashArray = "3", fillOpacity = 0.9,
    highlight = highlightOptions(
      weight = 5, color = "#666",
      dashArray = NULL, fillOpacity = 0.9,
      bringToFront = TRUE),
    label = labels_cases.county,
    layerId = ~id,
    labelOptions = labelOptions(
      style = list("font-weight" = "normal", padding = "3px 8px"),
      textsize = "15px", direction = "auto"),
    options = pathOptions(pane = "polygons")) %>%
  addLegend(data = counties1, pal = pal.county.quantile,
            values = ~Infect_risk, opacity = 0.7,
            title = NULL, position = "bottomright")
```

6.5.2   A County-level Map of COVID-19 Control Policy

This section presents another example of a county-level map of COVID-19 related control policies using the dataset IDDA::counties2 with the output shown in the bottom panel of Figure 6.7.

```
data(counties2)

pal.county.factor <- colorFactor(
  palette = c('#ef8a62', '#67a9cf'),
  domain = c('YES', 'NO')
)

# Draw the county-level map
dmap2a <- leaflet() %>%
  setView(-96, 37.8, zoom = 4) %>%
  addTiles() %>%
  # Add additional panes to control layer order
  # Display borders (zIndex: 420) above the polygons (zIndex: 410)
  addMapPane("polygons", zIndex = 410) %>%
  addMapPane("borders", zIndex = 420) %>%
  # Add state polygons (borders pane)
  addPolygons(
    data = states1, fill = FALSE, weight = 1,
    color = "gray", fillOpacity = 0,
    options = pathOptions(pane = "borders")
  ) %>%
  # Add county polygons (polygons pane)
  addPolygons(
    data = counties2,
    fillColor = ~pal.county.factor(X2020.05.01),
    weight = 1, opacity = 0.9, color = "gray",
    dashArray = "3", fillOpacity = 0.9,
    highlight = highlightOptions(
      weight = 5, color = "#666",
      dashArray = NULL, fillOpacity = 0.9,
      bringToFront = TRUE),
    label = labels_cases.county,
    layerId = ~id,
    labelOptions = labelOptions(
      style = list("font-weight" = "normal", padding = "3px 8px"),
      textsize = "15px", direction = "auto"),
    options = pathOptions(pane = "polygons")) %>%
  # Add legend
  addLegend(data = counties2, pal = pal.county.factor,
            values = ~(X2020.05.01), opacity = 0.7, title = NULL,
            position = "bottomright")
```

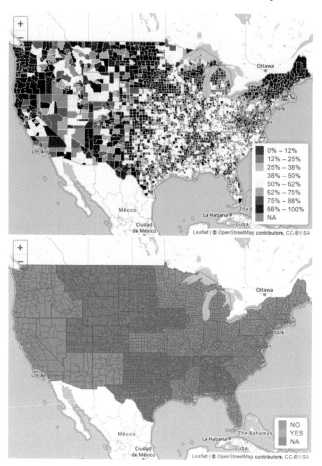

FIGURE 6.7: Two examples of the county-level map. Top: a county-level map of COVID-19 infection risk on December 31, 2020. Bottom: a county-level map of COVID-19 related control policy on May 1, 2020.

6.6 Spot Maps

A **spot map** is a map that depicts the geographic location of people who share a common characteristic, such as the number of infectious disease cases. Spot maps are typically used to visualize clusters or outbreaks with a small number of cases.

Next, we draw a spot map and highlight the top 10 counties with the largest cumulative infected count of COVID-19 on December 31, 2020.

First, let us prepare the data required.

```
library(IDDA)
data(features.county)
data(county.top10)

# Combine the datasets with useful variables
location.county <- features.county %>%
  dplyr:: select(ID, Longitude, Latitude)
county.top10.today <- county.top10 %>%
  select(ID, County, State, Infection)
names(county.top10.today)[4] <- 'Count'
df <- left_join(county.top10.today, location.county, key = "ID")
```

We start to draw a base map; see the top left panel of Figure 6.8.

```
dmap3 <- leaflet() %>%
  setView(-96, 37.8, zoom = 4) %>%
  addTiles()
```

6.6.1 Adding Circles

As in the top right panel of Figure 6.8, we can add circles to the map to highlight the top ten counties in the data using addCircles(). Circles and circle markers are similar, except circles have their radii specified in meters, while circle markers are specified in pixels. As a result, the size of circles will change as the user zooms in and out, while circle markers remain a constant size on the screen regardless of zoom level.

```
dmap3 <- dmap3 %>%
  addCircles(data = df, lng = ~Longitude, lat = ~Latitude, weight = 1,
             radius = ~sqrt(Count) * 200, popup = ~County)
```

```
dmap3
```

Each point can have text added to it using either a popup (appears only on click) or a label (appears either on hover or statically). We will describe the details below.

6.6.2   Adding Popups

Popups are small boxes that contain HTML outputs such as texts or hyperlinks and point to a specific point or location on the map.

Popups are frequently used so that they appear when markers or shapes are clicked. The Leaflet package's marker and shape functions take a popup argument, where we can pass in HTML commands to easily attach a simple popup.

For instance, we can label each county with the name of the county, state, and the reported cumulative infected cases.

```
labels_cases.county <- sprintf(
  "<strong>%s</strong>, <strong>%s</strong>
  <br/> Cum. Infected Cases on 2020-12-31: %g <br>",
  df$County, df$State, df$Count) %>%
  lapply(htmltools::HTML)
```

Using `lapply(htmltools::HTML)`, one can pass the information to leaflet, so it knows to treat each label as HTML instead of as plain text.

If we only want the information to appear when we click on the point, we should instead use `popup = ~labels_cases.county` like the following:

```
dmap3 %>%
  addMarkers(data = df, lng = ~Longitude, lat = ~Latitude,
             popup = ~labels_cases.county)
```

The resulting plot is shown in the bottom left panel of Figure 6.8.

6.6.3   Adding Labels

A **label** is textual or HTML content attached to markers and shapes and is visible at all times or when the mouse is moved over it. Unlike popups, we don't have to click a marker or polygon to see the label.

```
dmap3 %>%
  addMarkers(data = df, lng = ~Longitude, lat = ~Latitude,
             label = ~labels_cases.county)
```

The resulting plot is given by the bottom right panel of Figure 6.8.

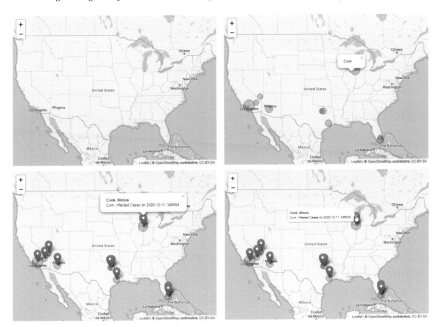

FIGURE 6.8: Examples of spot maps. Top left: drawing a base map. Top right: adding circles using `addCircles()`. Bottom left: adding popups using `lapply(htmltools::HTML)` and `addMarkers()`. Bottom right: adding labels using `addMarkers()`.

6.7 Integrating Leaflet with R Shiny

The "leaflet" package includes powerful and convenient features to integrate with Shiny applications. Most Shiny output widgets are integrated into an app by including a render function in the server function and an output for the widget in the UI definition. We can call the `leafletOutput()` function in the UI and assign a `renderLeaflet()` call to the output on the server side to return a "leaflet" map object for Leaflet maps.

```
shinyServer(function(input, output) {
  output$map <- renderLeaflet({})
})
```

server.R

```r
shinyServer(function(input, output) {
  # Prepare data
  location.county <- IDDA::features.county %>%
    dplyr:: select(ID, Longitude, Latitude)
  county.top10.today <- IDDA::county.top10 %>%
  select(ID, County, State, Infection)
  names(county.top10.today)[4] <- 'Count'
  df <- left_join(county.top10.today, location.county, key = "ID")
  # Prepare popup label
  labels_cases.county <- sprintf(
    "<strong>%s</strong>, <strong>%s</strong>
    <br/> Cum. Infected Cases on 2020-12-31: %g <br>",
    df$County, df$State,
    df$Count
  ) %>% lapply(htmltools::HTML)
  # Draw a spot map with popup
  output$map <- renderLeaflet({
    leaflet() %>%
      setView(-96, 37.8, zoom = 4) %>%
      addTiles() %>%
      addCircles(data = df, lng = ~Longitude, lat = ~Latitude,
                 weight = 1, radius = ~sqrt(Count)*200,
                 popup = ~County) %>%
      addMarkers(data = df, lng = ~Longitude, lat = ~Latitude,
                 popup = ~labels_cases.county)
  })
})
```

ui.R

```r
library(shiny)
library(dplyr)
library(leaflet)
library(IDDA)
shinyUI(
  # Use a fluid layout
  fluidPage(
    # Give the page a title
    titlePanel(
      HTML(
        paste("Top 10 counties with the largest cumulative", '<br/>',
              "infected count on December 31, 2020")
```

```
    )),
    mainPanel(leafletOutput("map"))
))
```

We can also run the following code to host this Shiny application using GitHub repository:

```
library(shiny)
runGitHub(repo = "IDDA_book", username = "FIRST-Data-Lab",
          subdir = "RShinyapp/leaflet_demo", ref = "main")
```

If everything is working correctly, we will see the application appear in the browser looking something like this:

FIGURE 6.9: The result of integrating leaflet with R Shiny.

6.8 Further Reading

The major resources for leaflet can be found on the following websites:

- https://rstudio.github.io/leaflet/;
- https://cran.r-project.org/web/packages/leaflet/index.html;
- https://geocompr.robinlovelace.net/adv-map.html;
- https://www.r-graph-gallery.com/bubble-map.

The following webpage provides more examples of Shiny with leaflet:

- `https://rstudio.github.io/leaflet/shiny.html;`

For a good reference to the spatial visualization for epidemiological analysis, see Chapter 3 of the following book:

- Pfeiffer, D., Robinson, T. P., Stevenson, M., Stevens, K. B., Rogers, D. J., Clements, A. C., et al. (2008). *Spatial analysis in epidemiology*, volume 142. Oxford University Press, Oxford.

6.9   Exercises

1. Create polygons based on the given coordinates:

   (a) Let

   ```
   x1 <- c(6, 8, 8, 6, 6)
   x2 <- c(6, 6, 4, 4, 6)
   ```

   (b) Let

   ```
   y1 <- c(5, 6, 8, 10, 5)
   y2 <- c(8, 3, 2, 8, 8)
   ```

   (c) Draw the two polygons in parts (a) and (b) on the same map.

2. The Washington Monument is located at longitude -77.0353 and latitude 38.8895.

   (a) Draw a base map and set the default view to and set zoom level 15.
   (b) Add a popup "Washington Monument" to your base map.
   (c) Add a label "Washington Monument" to your base map.

3. We will use the state-level COVID-19 data (`I.state`) available in the IDDA package, and the geospatial information from:

```
library(geojsonio)
urlRemote  <- "https://raw.githubusercontent.com/"
pathGithub <- "PublicaMundi/MappingAPI/master/data/geojson/"
fileName   <- "us-states.json"
geojson_read(x = paste0(urlRemote, pathGithub, fileName),
             what = "sp")
```

(a) Calculate the weekly risk for each state and create a
     weekly_risk variable, which contains the number of new in-
     fected cases in the week from December 25 to December 31,
     2020, divided by the population in the corresponding state.
(b) Draw a choropleth map to display the weekly risk for each state.
     You can change the opacity and the weight of the borderlines
     according to your aesthetic preferences. Use colorBin to color
     the states.
(c) Generate the highlighted label with the state name and the
     value of weekly_risk, and population, and display the label
     when the mouse moves over the state.
(d) Save your leaflet map as an HTML file.

4. We will use the county-level COVID-19 data (I.county) available
   in the IDDA package, and the geospatial information from:

```
urlRemote  <- "https://raw.githubusercontent.com/"
pathGithub <- "plotly/datasets/master/"
fileName   <- "geojson-counties-fips.json"
geojson_read(x = paste0(urlRemote, pathGithub, fileName),
             what = "sp")
```

(a) Calculate the weekly risk of each county and create a
     weekly_risk variable, which contains the number of new in-
     fected cases in the week from December 25 to December 31,
     2020, divided by the population in the corresponding county.
     The county population information is given in the pop.county
     dataset in the IDDA package.
(b) Draw a choropleth map to display the weekly risk for each
     county. You can change the opacity and the weight of the bor-
     derlines according to your aesthetic preferences. Use colorQuan-
     tile to color the counties.

(c) Generate the highlighted label with the county, state name and the value of `weekly_risk`, and population, and display the label when the mouse moves over the county.

(d) Draw a spot map and highlight the top 10 counties with the highest weekly risk in the week from December 25 to December 31, 2020.

(e) Save each of the above leaflet maps as an HTML file.

7

Epidemic Modeling

Epidemic modeling is a vast field. The epidemic models developed have played important roles in the investigation of infectious diseases. The primary goals of the epidemic models are (1) to obtain a better understanding of how the virus spreads; (2) to discover which variables contribute to the transmission of the epidemic; and (3) to aid in decision-making regarding an effective control plan for diseases. Mechanistic models, phenomenological models, and hybrid models are all sorts of epidemiological modeling methodologies. Although some of these strategies are complementary, they begin at various places with different details and features. This chapter provides a quick overview of several epidemic modeling techniques useful in epidemiology without providing a detailed background for each of them. The end of the chapter also provides a glossary of terms in epidemic modeling with their definitions to help data scientists without a background in epidemiology.

This chapter is by no means exhaustive, and inevitably there are many other types of epidemic models that we have excluded. We hope this chapter will prompt readers to study more to further extend the knowledge of epidemic modeling.

7.1   An Introduction to Epidemic Modeling

Daniel Bernoulli is credited with developing the first disease transmission model (Dietz and Heesterbeek, 2002). Bernoulli developed an epidemiological model in 1760 and used it to demonstrate that life expectancy increased due to vaccination against smallpox. When William Farr noticed the similarity between the time series of smallpox incidence and the normal distribution, he was the first to calibrate (or fit) a model of disease transmission. In 1889, En'ko investigated heterogeneity in measles transmission, using a discrete chain binomial model for the spread of infection in a susceptible population; see En'Ko (1989). Kermack and McKendrick used differential equations to build the foundation of the modern susceptible-infected-recovered mass-action model in 1927, expanding on existing models; see Kermack and McKendrick (1927).

These and other models have served as the foundation for modern disease modeling, which has proven to be an invaluable tool in advancing our understanding of best practices for infectious disease control. Epidemic models not only enable the evaluation of public health interventions (e.g., screening programs, vaccines, and projection of future disease burden in various contexts), but also help understand the biological transmission mechanisms and answer fundamental questions such as why some people are uninfected during an outbreak.

Three main types of models are considered in epidemic modeling, including (i) mechanistic models, (ii) phenomenological models, and (iii) hybrid models. In the following, we will give a brief introduction to each type of model. More in-depth studies will be covered in Chapters 8–12.

7.2 Mechanistic Models

Mechanistic models aim to explicitly account for the mechanisms of interactions among system components. The biological mechanisms that drive infection dynamics are hypothesized explicitly in these models.

7.2.1 Compartment Modeling

A compartment model usually employs a collection of mathematical equations to describe how individuals are transferred from one compartment to another. The Susceptible-infected-removed (SIR) model Kermack and McKendrick (1927) and its variants with additional compartments are commonly considered to conceptualize the dynamics of representative populations in epidemiology.

The traditional SIR model consists of three compartments of a population: the susceptible compartment, the infected compartment, and the removed compartment. The susceptible compartment (S) represents those who are susceptible to infection, the infected compartment (I) indicates people who are infectious, and the removed compartment (R) represents those who have permanent infection-acquired immunity. Due to its simplicity and a small number of parameters, the SIR model has gained popularity when describing the trajectory of diseases. For example, Miranda et al. (2019) calibrated the parameters of the SIR model for each season separately using the Nelder-Mead optimization algorithm to predict weekly influenza-like-illness (ILI) incidence. Chen et al. (2021) employed the SIR model for Canadian COVID-19 data. However, it has been known that the SIR model is often too simple to represent the complex processes of disease.

If an exposed period (infected but not yet infectious) is significant for a certain disease, its absence in the model would be related to poor predictions Brauer (2008). The SEIR model takes into account such latent period using an exposed (E) compartment. The SEIR has been applied to malaria transmission by formulating a system of equations for both human and mosquito populations Mojeeb et al. (2017) and Ebola hemorrhagic fever by considering vaccination as a control strategy (Durojaye and Ajie, 2017). Roda et al. (2020) considered both SIR and SEIR models and predicted the COVID-19 epidemic. Although a Poisson or negative binomial distribution is commonly assumed for count data, its probability model was approximated by normal distribution as newly confirmed cases grow quickly. The affine invariant ensemble Markov Chain Monte Carlo algorithm was implemented for posterior distributions of model parameters using uniform priors for model parameters to reduce the impact of nonidentifiability in the model.

In Chapter 8, we will study compartment models with a more detailed structure and see how they can help study infectious diseases.

7.2.2 Agent-based Methods

Individual-based models, or agent-based models, are a powerful simulation modeling technique based on a collection of agents who are interacting with autonomous decision-making entities. In epidemic modeling, agents could take actions appropriate for the representative states, such as susceptibility, infection, or recovery. Interactions among those agents within a specific environment characterized by a social contact network are a feature of agent-based modeling. In addition, agents may move from one state to another while they may evolve within a state. A typical agent-based model consists of three key elements: (i) a population-based on the studied population's demographic characteristics, (ii) a social contact network among the population's agents (individuals), and (iii) a disease model that translates the edge weights in the social contact network into infection probability (Hoertel et al., 2020). Agent-based models can exhibit complex behavior patterns through simulations of simple agents' interactions.

A good example of using this approach for COVID-19 is by a team from Imperial College London (Ferguson et al., 2020), who applied an agent-based modeling approach to the UK data. They included several non-pharmaceutical interventions when forecasting the spread of COVID-19. Hoertel et al. (2020) considered a stochastic agent-based model to predict the cases of COVID-19 in France. The Global Epidemic and Mobility (GLEAM[1]) project was launched

---

[1]https://covid19.gleamproject.org/

to analyze the spatiotemporal spread of COVID-19 in the continental US using an agent-based model.

Agent-based models make use of further detailed information on disease states, such as individual patterns and behaviors, which are not easily characterized from a compartmental form. They allow for capturing the relationships among individuals and representing heterogeneity in their attributes (Rahmandad and Sterman, 2008). Besides, agent-based models are useful to model experiments that may be unethical or impossible due to several issues in the real world.

Agent-based models require intensive computational burdens, constraining sensitivity analysis. Because data on contact networks and the distribution of individual attributes is difficult to come by and highly uncertain, extensive sensitivity analysis is required to ensure reliable results.

7.3 Phenomenological Models

A phenomenological model is usually a statistical model that characterizes and forecasts epidemics' observed effects without incorporating biological mechanisms and postulating conjectures that explain the observed phenomena.

7.3.1 Time Series Analysis

Time series analysis focuses on capturing the underlying patterns and forecasting the future values by employing the past observations of a random variable. The two most widely used approaches in time series forecasting are exponential smoothing (ETS) and autoregressive integrated moving average (ARIMA) models. ETS models are used to describe the trend and seasonality of time series data, whereas ARIMA models are used to describe the autocorrelations. Below, we will give a short introduction to ETS and ARIMA models. Chapter 9 introduces the details of time series analysis and how to make the epidemic prediction using R.

7.3.1.1 Exponential Smoothing

The exponential smoothing method was proposed by Brown (1959), Holt (1957), and Winters (1960) in the late 1950s. Traditional moving average methods use equal weights for previous observations, whereas exponential smoothing uses exponential functions to place smaller weights for observations in the further past.

Since its framework usually can generate reliable forecasts quickly, the simple exponential smoothing method is suitable for forecasting data with no clear trend or seasonal pattern.

- **Holt's Linear Trend Method**: Holt (1957) proposed a linear trend method based on simple exponential smoothing to allow data with a trend to be predicted. This method involves three equations, including (i) a forecast equation, (ii) a smoothing equation for the level, and (iii) a smoothing equation for the trend. As a result, the h-step-ahead forecast equals the sum of the most recent estimated level and h times the most recent estimated trend value, and the forecasts are linear to h. The long-term forecast is usually indefinitely increasing or decreasing due to the constant trend estimated by Holt's method.

- **Damped Trend Method**: Because constant trend has its limitations, Gardner and Mckenzie (1985) proposed a "damped" method, which includes a parameter that "dampens" the trend to a flat line in the future. As pointed out in Hyndman and Athanasopoulos (2018), the "damped" method is one of the most popular time-series methods for forecasting.

- **Holt-Winters' Seasonal Method**: Holt (1957) and Winters (1960) extended Holt's method to the Holt-Winters' Seasonal Method, which can analyze time series with seasonality by taking into account a systematic trend or a seasonal component. The Holt-Winters seasonal method adds one more smoothing equation for the seasonal component to Holt's Linear Trend Method's three equations.

7.3.1.2 AutoRegressive Integrated Moving Average Models

The AutoRegressive Integrated Moving Average (ARIMA) model provides another approach to time series forecasting. Combining the autoregression (AR) and moving average (MA) models, the ARIMA model exploits the dependencies among the time series to extract local patterns and remove high-frequency noise from the data.

High interpretability is one of the benefits of the ARIMA model. Thus, researchers are able to gain a deep understanding of the relationship between the current and the past situations and explore the influence of some exogenous variables. In addition, the ARIMA has a high accommodative ability through the simple updates for a model using recent events. However, the ARIMA model is not suitable to deal with nonlinear patterns or relationships.

7.3.2   Regression Methods

Regression is one of the popular methods when estimating future prevalence or targets of interest in epidemiology. For example, Altieri et al. (2021) considered five regression models with different trends in COVID-19 death counts. Another example of regression methods is the "COVID-19 Forecasts using Fast Evaluations and Estimation" (COFFEE) proposed by the Los Alamos National Laboratory; see Castro1 et al. (2020). We will introduce (generalized) linear regression models and other regression models in Chapter 10. We will also show how regression and discrimination analysis can be used to quantify the effect of a set of explanatory variables on the spatial distribution of a specific outcome.

7.3.3   Machine Learning Methods

With their great flexibility in capturing disease spread patterns, machine learning methods are widely used in infectious disease modeling and prediction. In the prediction of COVID-19, there are two major categories of machine learning methods. The first type of method uses epidemic models trained using machine learning algorithms. Zou et al. (2020) proposed a variant of the SEIR model that accounts for COVID-19 cases that have not been tested or reported, and the model is estimated using a standard gradient-based optimizer. Arik et al. (2020) integrated machine learning into compartmental disease modeling. For better model estimation with limited training data, learning mechanisms such as masked supervision from partial observations and partial teacher-forcing is considered to minimize error propagation. The second kind of method is based on conventional machine learning methods. Sujath et al. (2020) applied linear regression, multilayer perceptron, and vector autoregression methods for predicting COVID-19 cases in India gathered from the web site of Kaggle. In Arora et al. (2020), recurrent neural network (RNN)-based long short-term memory (LSTM) was used to predict the number of COVID-19 reported cases for the state-level data in India. Chapter 11 provides some implementation details of neural network models for forecasting.

7.4   Hybrid Models and Ensemble Methods

Because of a lack of understanding of the mechanical details, hybrid modeling often refers to methods where part of a model is formulated based on mechanical principles, and part of the model should be inferred from data. In the past two decades, statistical and machine learning-based time series approaches have been considered when exploring the dynamics of infectious diseases, for example, seasonal influenza. The application of hybrid and ensemble methodologies has become more visible and attractive in infectious disease forecasting. Ensemble models combine multiple forecasting algorithms for better prediction performance by reducing the instability of the forecast, especially when there is much uncertainty to find the best model. In COVID-19 studies, a growing number of hybrid methods have emerged by combining features of mechanistic and phenomenological models. Examples of hybrid models will be given in Chapter 12. We will also introduce ensemble methods using multiple forecasting algorithms to improve the predictive performance in Chapter 12.

7.5 Epidemic Modeling: Mathematical and Statistical Perspectives

Epidemic modeling has three main aims (Daley and Gani, 2001): (i) to understand better the mechanisms by which diseases spread; (ii) to identify which factors contribute to the spread of the epidemic, and therefore how we may control it; (iii) to predict the future course of the epidemic. Although there are many epidemic modeling methods, mathematical and statistical models have played important roles in COVID-19 studies. As demonstrated in Figure 7.1, mathematical and statistical approaches are complementary, but their starting points are different, and the corresponding models tend to incorporate different details.

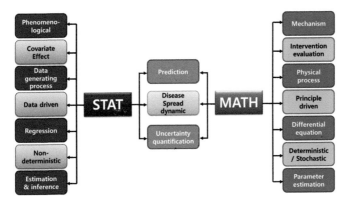

FIGURE 7.1: Mathematical and statistical perspectives on epidemic modeling.

As mentioned above, the fundamental concept of infectious disease epidemiology is investigating how the diseases spread. Mathematical models are undeniably useful in understanding infectious disease spread dynamics (e.g., when the peak will occur and whether resurgence will occur) and the effects of control measures (Keeling and Rohani, 2008). One essential type of mathematical models is the class of mechanistic models such as the Susceptible - Infectious - Removed (SIR) compartmental model or the Susceptible - Exposed - Infectious-Recovered model (SEIR) as illustrated in Figure 7.2; see details in Brauer et al. (2008) and Lawson et al. (2016). Mechanistic models express explicit hypotheses about the biological mechanisms that drive infection dynamics, and they work well when evaluating the efficacy of hypothetical Nonpharmaceutical Interventions (NPIs) in preventing disease spread (Lessler and Cummings, 2016).

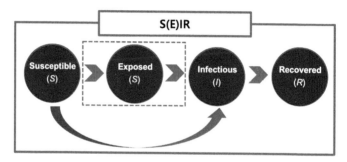

FIGURE 7.2: An illustration of SIR and SEIR models.

Statistical modeling has presented many successes within the scientific field in analyzing data and compiling information about the mechanisms producing the data. Statistical modeling is a powerful tool that can be utilized by extracting information about disease spread in epidemic studies (Held et al., 2020). There are two cultures in statistical modeling (Breiman, 2001): the data modeling culture and the algorithmic modeling culture. The first assumes that the data are generated by a given stochastic data model, which is typically designed for inference about variable relationships as well as prediction. Algorithmic models are designed to make the most accurate predictions possible by treating the data mechanism as unknown.

Other factors, such as demographic characteristics, socioeconomic status, and control policies may also be responsible for temporal or spatial patterns in the spread of infectious diseases. The disease's spread, for example, varies greatly across different geographical regions. Local features, such as socioeconomic factors and demographic conditions, can significantly affect the epidemic's trajectory. These data are usually supplemented with the population information at the county level. Moreover, the capacity of the health care system and control measures also have a significant impact on the spread of the epidemic. Regression is a widely used statistical modeling method in epidemic studies because it produces a combination of the variables with weights indicating the impact of the variable (Jewell, 2003). It can assist us in determining which factors matter most and how those factors interact with each other. Another benefit of regression analysis is that it can be used to understand various patterns in data and provide valuable insights in understanding which factors contribute to the spread of COVID-19.

It is important to predict the spread speed and severity of COVID-19 to manage resources, develop strategies to deal with the epidemic, and ultimately assist in prevention efforts. Mathematical models are able to mimic how the disease gets spread and can be used to project or simulate future transmission scenarios under various assumptions. Statistical models are more oriented towards predictions (Held et al., 2020). In fact, predictions play a critical role in statistical modeling. Time series analysis, for example, is a statistical forecasting technique that uses a series of historical observations to extrapolate patterns into the future. Machine learning makes predictions based on previously learned properties from training data. However, models that are purely statistical often do not account for the circumstance of transmission, and so they exclusively depict the observed data. They provide little information about the actual mechanism. Therefore, long-term predictions generally do not favor purely statistical models. The ability to quantify the uncertainty in the prediction, especially at the early phase of an epidemic where there is limited data, is another advantage of statistical modeling. For example, statistical models can provide a prediction interval to understand the uncertainty surrounding the forecast (Brockwell and Davis, 2016).

What can be drawn from the above is that mathematical models are usually constructed in a more principle-driven manner, while statistical models are more data-driven. Although both mathematical and statistical models can be used to study the effect of NPIs and make predictions, the implementation details are different, and an understanding of the corresponding limitations is crucial. Researchers that work to advance epidemic modeling will need to appreciate and utilize the complementary strengths of mathematical and statistical models for maximum efficiency.

7.6 Some Terms in Epidemic Modeling

In this section, we introduce some terms often used in infectious disease modeling.

- **Incidence:** The number of new cases in a given period expressed as a percentage of people infected per year (cumulative incidence) or number per person-time of observation (incidence density).

– Example: Meningococcal disease epidemics are common in Auckland, New Zealand, with annual incidences of up to 16.9 per 100,000 people.

- **Prevalence:** The number of cases at a given time expressed as a percentage of the total number of cases at that time.

  – Example: Obesity prevalence was 12.8 percent in a group of children aged 3 to 4 years old, according to a recent Scottish study.

- **Attack rate:** The proportion of non-immune exposed individuals who become clinically ill.

  – Example: In an outbreak of gastroenteritis with 50 cases among a population at risk of 2500, the attack rate of disease is $50/2500 = 0.02$.

- **Primary/secondary cases:** A primary case is a person who enters a population and infects it. Secondary cases are those who contract the infection afterward. "Waves" or "generations" are terms used to describe the spread of the virus.

- **Case fatality:** The proportion of infected people who die as a result of the infection.

- **Mortality:** The proportion of the population that dies each year as a result of the disease, calculated as the product of the incidence and case fatality rate.

  – Example: Consider two populations. There are 1,000 people in Population 1; 300 of them have the disease, and 100 of them die from it. The disease has a mortality rate of $100/1,000$ or 10%. The fatality rate in this case is $100/300 = 33\%$. There are 1,000 people in Population 2, 50 of whom have the disease and 40 of whom have died as a result of it. The mortality rate is $40/1,000$, or 4%; however, the case fatality rate is $40/50$, or 80%.

- **Reproductive rate:** The potential of an infectious disease spreading. Factors that have an impact include:

  – the likelihood of infection spreading from an infected person to a susceptible person;
  – the contact frequency with the population;
  – the duration of infection;
  – the proportion of people who are immune.

- **Transmission routes:** The pathway of causative agents from a source to infection of a susceptible host, which can be direct (mucous membrane to mucous membrane, cross placental, blood or tissue, skin to skin, sneezes or coughs) and indirect (water, air borne, food borne, vectors or objects).

- **Reservoir:** The population of organisms or the specific environment in which an infectious pathogen naturally lives and reproduces, or upon which the pathogen primarily depends for its survival.

- **Incubation period:** The time it takes for a disease to develop after being exposed to it (the time between inoculation and symptom expression).

- **Infectious period:** The length of time a person can transmit disease.

- **Latent period:** The period of infection without being infectious (the time between inoculation and infectiousness of the host). This may occur right after exposure or late in the disease.

- **Epidemic:** The occurrence of cases of illness that lasts longer than expected.
  - **Endemic:** An epidemic whose incidence has remained stable for a long time.
  - **Pandemic:** A worldwide outbreak.

7.7 Further Reading

There are a few books available that cover the use of scientific models in epidemiology:

- Becker, N. G. (2017). *Analysis of infectious disease data.* Chapman and Hall/CRC.
- Bjørnstad, O. N. (2018). *Epidemics: Models and data using R.* Springer.
- Brauer, F., Castillo-Chavez, C. and Feng, Z. (2019). *Mathematical models in epidemiology* (Vol. 32). New York: Springer.
- Broemeling, L. D. (2021). *Bayesian analysis of infectious diseases: COVID-19 and beyond.* Chapman and Hall/CRC.
- Held, L., Hens, N., D O'Neill, P. and Wallinga, J. (Eds.). (2019). *Handbook of infectious disease data analysis.* CRC Press.
- Jewell, N. P. (2003). *Statistics for epidemiology.* Chapman and Hall/CRC;
- Lawson, A. B., Banerjee, S., Haining, R. P. and Ugarte, M. D. (Eds.). (2016). *Handbook of spatial epidemiology.* CRC Press.

8

Compartment Models

Infectious disease dynamics mathematical models have a long history, dating back over a century. Simple mathematical formulations describing individuals' transition in a population between "compartments" that capture their infection status yield surprisingly significant insight. Their elegance and simplicity make it simple to expand to more complexities by adding compartments.

This chapter presents some classical compartment models, such as the SIR and SEIR models, and their characteristics are also discussed. We formulate our descriptions of disease transmission as compartmental models. The population under study is divided into compartments: susceptible, infected but not yet infectious, infectious, and recovered compartments, for example. We illustrate basic simulations to look at how each compartment behaves over time and describe how to numerically estimate the parameters, such as the rate of infection for the compartment model. In addition, this chapter demonstrates how to use compartment models in the R environment to predict the course of an epidemic by calculating associated parameters between compartments, such as the rate of infection or recovery. The final part of this chapter introduces the concept of the basic reproduction number, the effective reproduction number, and herd immunity.

A single chapter in a book like this one cannot do justice to the richness of the compartment models. Therefore, in this chapter, we just focus on a few selected models or topics that seem to be of paramount importance or provide a good starting point when it comes to the use of compartment models.

8.1 SIS Models

Because many diseases do not confer immunity, infectives revert to the susceptible class after recovery. Most diseases transmitted by bacterial or helminth agents, as well as most sexually transmitted diseases, can be studied using these models (including gonorrhea, but not such diseases as AIDS, from which there is no recovery).

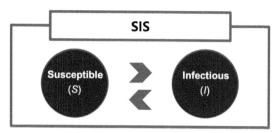

FIGURE 8.1: A simple SIS model.

In an SIS model, the total population size $N = S(t) + I(t)$. The simplest SIS model, due to Kermack and McKendrick (1927), is

$$
\begin{aligned}
\frac{dS(t)}{dt} &= -\beta I(t)\frac{S(t)}{N} + \gamma I(t), \\
\frac{dI(t)}{dt} &= \beta I(t)\frac{S(t)}{N} - \gamma I(t),
\end{aligned}
$$

where γ is the rate of infected individuals who re-enter the susceptible class, β is the effective contact rate, and

$$
\beta \propto \left(\frac{\text{infection}}{\text{contact}}\right) \times \left(\frac{\text{contact}}{\text{time}}\right).
$$

The simplicity of the SIS model makes it easy to compute and understand the trajectory of the disease. However, the SIS model also makes several assumptions about the population. It assumes (i) the rate of new infections is determined by the incidence of mass action; (ii) infectives leave the infective class and return to the susceptible class at a rate of γ; (iii) no one enters or leaves the population; (iv) no disease deaths occur, and the total population remains constant at N.

8.2   SIR Models

Consider the SIR model Kermack and McKendrick (1927) in a fixed population of size $N = S(t) + I(t) + R(t)$:

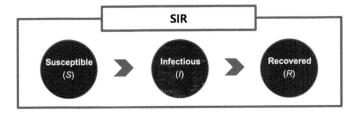

FIGURE 8.2: An SIR model.

$$\frac{dS(t)}{dt} = -\beta I(t)\frac{S(t)}{N},$$
$$\frac{dI(t)}{dt} = \beta I(t)\frac{S(t)}{N} - \gamma I(t),$$
$$\frac{dR(t)}{dt} = \gamma I(t),$$

where β is the effective contact rate, and

$$\beta \propto \left(\frac{\text{infection}}{\text{contact}}\right) \times \left(\frac{\text{contact}}{\text{time}}\right),$$

i.e., (probability of transmission given a contact between a susceptible and an infectious individual) × (average rate of contact between susceptible and infected individuals); γ is the removal rate, and γ^{-1} is the average infectious period. The logic of the transmission term is that β is the contact rate among hosts times the probability of infection given a contact.

Let $s(t) = S(t)/N$, $i(t) = I(t)/N$ and $r(t) = R(t)/N$. Dividing the equations for $S(t)$, $I(t)$ and $R(t)$ by N we get the deterministic SIR epidemic model for this process in the form:

$$\frac{ds(t)}{dt} = -\beta i(t)s(t),$$
$$\frac{di(t)}{dt} = \beta i(t)s(t) - \gamma i(t),$$
$$\frac{dr(t)}{dt} = \gamma i(t).$$

The SIR model also makes several assumptions about the population. It assumes that (i) constant (closed) population size: N; (ii) constant rates (e.g., transmission, removal rates); (iii) no demography (i.e., births and deaths); (iv) well-mixed population, meaning that any infected person has a probability of contacting any susceptible person that is reasonably close to the average.

The "deSolve" package in R provides the `ode()` function, a general solver for ordinary differential equations (ODEs). First, we define a function that computes the derivatives of each compartment at `time` for an SIR model.

```
## Load the deSolve package
library(deSolve)
## Create an SIR function
sir <- function(time, state, parameters) {
  with(as.list(c(state, parameters)), {
    dS <- -beta * S * I
    dI <-  beta * S * I - gamma * I
    dR <- gamma * I
    return(list(c(dS, dI, dR)))
  })
}
```

Using the function with information on derivatives, the `ode()` function returns the trajectories of each compartment by solving the ODE. Below, we consider a simulation example with $s(0) = 0.999$, $i(0) = 0.001$, $r(0) = 0.0$, $\beta = 0.3$ and $\gamma = 0.1$.

```
# Set parameters
# Proportion in each compartment: Susceptible 0.999,
# Infected 0.001, Recovered 0
init <- c(S = 0.999, I = 0.001, R = 0.0)
# beta: infection parameter; gamma: recovery parameter
parameters <- c(beta = 0.3, gamma = 0.1)
# Time frame
times <- seq(0, 300, by = 1)

# Solve using ode
# (General Solver for Ordinary Differential Equations)
out.sir <- ode(y = init, times = times, func = sir, parms = parameters)
# Change to data frame
out.sir <- as.data.frame(out.sir)
# Show the data
head(out.sir, 5)
```

```
##   time      S        I         R
## 1    0 0.9990 0.001000 0.0000000
## 2    1 0.9987 0.001221 0.0001109
```

```
## 3    2 0.9983 0.001492 0.0002464
## 4    3 0.9978 0.001821 0.0004116
## 5    4 0.9972 0.002223 0.0006131
```

The development of the prevalence rate in the SIR model is illustrated in the top panel of Figure 8.4. The solid curve corresponds to the infectious compartment, $I(t)$, the dashed curve to the susceptible compartment, $S(t)$, and the dotted curve to the recovered compartment, $R(t)$. It is worth noting that $S(t)$ enters $I(t)$ without returning to $S(t)$, implying that the $S(t)$ curve has reduced over time t. Furthermore, $I(t)$ has a bell-shaped curve because its increase proportional to $S(t)$ is very high at the time of the disease outbreak, but the transition from $I(t)$ to $R(t)$ is raised at the end time. The recovery compartment $R(t)$ exhibits a steady growth over time owing to not returning to the preceding compartment.

8.3   SIR Models with Births and Deaths

Because the time scale of an epidemic is generally much shorter than the demographic time scale, we have omitted births and deaths from our description of epidemic models. In effect, we have used a time scale with a negligible number of births and deaths per unit of time. There are, however, diseases that are endemic in many parts of the world and kill millions of people every year.

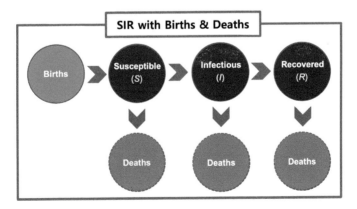

FIGURE 8.3: An SIR model with birth and death.

We consider the following SIR model with births and deaths:

$$
\begin{aligned}
\frac{dS(t)}{dt} &= \mu\{N - S(t)\} - \beta I(t)\frac{S(t)}{N}, \\
\frac{dI(t)}{dt} &= \beta I(t)\frac{S(t)}{N} - (\mu + \gamma)I(t), \\
\frac{dR(t)}{dt} &= \gamma I(t) - \mu R(t),
\end{aligned}
$$

where β is the effective contact rate, and

$$
\beta \propto \left(\frac{\text{infection}}{\text{contact}}\right) \times \left(\frac{\text{contact}}{\text{time}}\right),
$$

i.e., (probability of transmission given a contact between a susceptible and an infectious individual) × (average rate of contact between susceptible and infected individuals). In addition, birth rate and death rate are equal, which is denoted as μ, and γ is the removal rate, and γ^{-1} is the average infectious period.

8.4 SEIR Models

Many diseases have a latent period in which the person is infected but not yet infectious. Such latent delay can be reflected by incorporating the additional exposed $(E(t))$ compartment in the SIR model. Consider the SEIR model in a population of size N, and note that $N = S(t) + E(t) + I(t) + R(t)$. In this case, we may also assume that people acquire resistance to the disease after recovering, but subsequently lose their immunity and re-enter the susceptible class at a rate of ω. Then, the SEIR model with demography (birth and death) and immunity can be characterized as follows:

$$
\begin{aligned}
\frac{dS(t)}{dt} &= \mu\{N - S(t)\} - \beta I(t)\frac{S(t)}{N} + \omega R, \\
\frac{dE(t)}{dt} &= \beta I(t)\frac{S(t)}{N} - (\mu + \sigma)E(t), \\
\frac{dI(t)}{dt} &= \sigma E(t) - (\mu + \gamma)I(t), \\
\frac{dR(t)}{dt} &= \gamma I(t) - (\mu + \omega)R(t),
\end{aligned}
$$

where ω^{-1} is the average duration of immunity, and σ^{-1} is the average latent period.

Note that, if we do not consider demography ($\mu = 0$) and immunity ($\omega = 0$), it becomes the simple SEIR model. We can solve a system of the ODE for an SEIR model using the ode() R function from the "deSolve" package. We need to define a function that computes the derivatives of each compartment at time for an SEIR model. For the initial value for each compartment in the SEIR, we assign $s(0) = 0.999$, $e(0) = 0.001$, $i(0) = 0.0$, $r(0) = 0.0$, $\beta = 0.3$, $\sigma = 0.25$ and $\gamma = 0.1$.

```r
seir <- function(time, state, parameters){
  with(as.list(c(state, parameters)), {
    dS <- -(beta * S * I)
    dE <- (beta * S * I) - sigma * E
    dI <- sigma * E - gamma * I
    dR <- gamma * I
    return(list(c(dS, dE, dI, dR)))
  })
}


init <- c(S = 0.999, E = 0.001, I = 0.0, R = 0.0)
parameters <- c(beta = 0.3, sigma = 0.25, gamma = 0.1)
times <- seq(0, 300, by = 1)
out.seir <- ode(y = init, times = times,
                func = seir, parms = parameters)
out.seir <- as.data.frame(out.seir)
head(out.seir, 5)
```

```
##    time      S         E          I         R
## 1     0 0.9990 0.0010000 0.0000000 0.000e+00
## 2     1 0.9990 0.0008093 0.0002130 1.115e-05
## 3     2 0.9989 0.0007100 0.0003715 4.079e-05
## 4     3 0.9987 0.0006696 0.0004992 8.451e-05
## 5     4 0.9986 0.0006693 0.0006103 1.401e-04
```

The bottom panel of Figure 8.4 depicts the evolution of the prevalence rate in the SEIR model mentioned above. The solid curve denotes the infectious compartment, $I(t)$, the long-dashed curve represents the exposed compartment, $E(t)$, the dashed curve represents the susceptible compartment, $S(t)$, and the dotted curve represents the recovered compartment, $R(t)$. Note that $S(t)$, $I(t)$, and $R(t)$ exhibit a pattern similar to the SIR simulation. Also, it has been discovered that $E(t)$ has a bell-shaped curve for the same reason as $I(t)$ does.

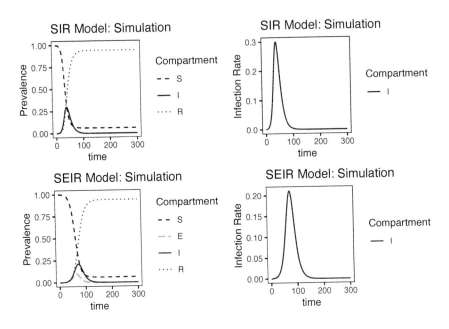

FIGURE 8.4: Simulation example for compartment model. Top left: prevalence of compartments in the SIR model. Top right: infection rate in the SIR model. Bottom left: prevalence of the SEIR model. Bottom right: infection rate in the SEIR model.

8.5 Parameter Estimation for Compartment Models

In this section, we discuss how to estimate the parameters of the SIR model without demography from the observed infectious cases. However, extension to other compartment models is straightforward by changing model parameters. Two methods are mainly considered to estimate the model parameters: (i) least-squares, and (ii) maximum likelihood methods.

8.5.1 Least-squares Method

The least-squares method allows us to quantify the gap between the data and the model's predictions. Then we can search through all of the possible values of a model's parameters to find the ones that minimize the discrepancy. For a given (β, γ), we denote $I(t; \beta, \gamma) = I(t)$. Then, the estimates can be obtained by minimizing the sum of squares of the observation and its fitted value

$$\sum_{j=1}^{n} \{I(t_j; \beta, \gamma) - Y_j\}^2,$$

where Y_j is the number of observed infected people at time t_j.

8.5.2 Maximum Likelihood Method

Let Y_j be the observed number of infected cases. We can assume

$$Y_j \sim \text{Poisson}\{pI(t_j; \beta, \gamma)\},$$

where the parameter p reflects a combination of sampling efficiency and the detectability of infections. Then, we can estimate parameters by maximizing the Poisson likelihood

$$\sum_{j=1}^{n} \left\{ -pI(t_j; \beta, \gamma) + Y_j \log(pI(t_j; \beta, \gamma)) - \log(Y_j!) \right\}.$$

If the errors $Y_j - I(t_j; \beta, \gamma)$ follow a normal distribution with constant variance σ^2, the parameters can be estimated by maximizing the normal likelihood

$$\sum_{j=1}^{n} \left\{ -\frac{1}{2} \log(2\pi\sigma^2) - \frac{(Y_j - I(t_j; \beta, \gamma))^2}{2\sigma^2} \right\}.$$

8.6 Implementation of Parameter Estimation in R

Below, we provide example codes for implementing the SIR to model (i) the number of patients suffering from an influenza-like illness (ILI) in the US from either the FluView portal[1] or the R package "cdcfluview"; and (ii) the spread of COVID-19 in Los Angeles, California, from the R package IDDA.

8.6.1 An Application to Influenza-like Illness Data

We mainly show how to use R to find the parameters in the SIR model by minimizing the sum of squares errors and by maximizing the likelihood function in this example. First, we load the number of patients suffering from ILI for

---

[1] https://gis.cdc.gov/grasp/fluview/fluportaldashboard.html

a certain region, such as national, hhs, census, or state, using the ilinet() function from the R package "cdcfluview". One chooses years for the CDC flu season of interest; for example, years = 2015 indicates the CDC flu season 2015–2016. Figure 8.5 shows Mid-Atlantic, New England, and Pacific regions from weeks 1 to 24 for the CDC flu season 2016.

```
library(dplyr)
library(plotly)
library(IDDA)
library(cdcfluview)

ili.data.raw = ilinet(region = "census", years = 2015)
ili.data = ili.data.raw %>%
  select(region, week, ilitotal) %>%
  filter((week > 0) & (week < 25) &
         (region %in% c("Mid-Atlantic", "New England", "Pacific")))



plot_ly() %>%
  add_trace(data = ili.data %>% filter(region == "Mid-Atlantic"),
            x = ~week, y = ~ilitotal, type = "scatter",
            name = "Mid-Atlantic", mode = "line",
            line = list(dash = 'dash',color = "#b2182b")) %>%
  add_trace(data = ili.data %>% filter(region == "New England"),
            x = ~week, y = ~ilitotal, type = "scatter",
            name = "New England", mode = "line",
            line = list(dash = 'dot', color = "#2166ac")) %>%
  add_trace(data = ili.data %>% filter(region == "Pacific"),
            x = ~week, y = ~ilitotal, type = "scatter",
            name = "Pacific", mode = "line",
            line = list(color = "#fddbc7"))
```

8.6.1.1   Least-squares Method

The R package "pomp" provides facilities for dealing with deterministic dynamics, which is an important special case.

```
library(pomp)
R.ili.data <- pomp(
```

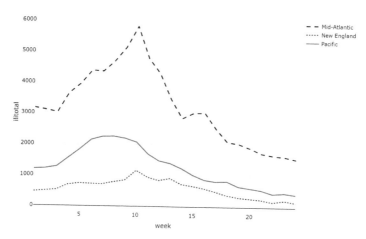

FIGURE 8.5: The number of patients suffering from ILI for the Mid-Atlantic, New England, and Pacific regions from weeks 1 to 24 for the CDC flu season 2016.

```
data = subset(ili.data, region == "New England", select = - region),
times = "week", t0 = 0,
skeleton = vectorfield(
  Csnippet("
    DS = - Beta * S * I / N;
    DI = Beta * S * I / N - gamma * I;
    DR = gamma * I;")),
rinit = Csnippet("
  S = S_0;
  I = I_0;
  R = N - S_0 - I_0;"),
statenames = c("S", "I", "R"),
paramnames = c("Beta", "gamma", "N", "S_0", "I_0"))
```

If we assume all the other parameters are known, one simple option to find optimal β would be grid searching a value that minimizes the sum of the squared error (SSE), which is the difference between the fitted infection curve with given β compared to the true observations. For example, it is well known that the infectious period of measles is about two weeks, so we can safely assume that $\gamma = 1$.

```
sse <- function(params) {
  x <- trajectory(R.ili.data, params = params)
```

```
  discrep <- x@states["I", ] - obs(R.ili.data)
  sum(discrep^2)
}
# Assume gamma = 1 is known
# Grid search for beta
f1 <- function(beta) {
  params <- c(Beta = beta, gamma = 1, N = 14000000,
              S_0 = 50000, I_0 = 450)
  sse(params)
}
beta <- seq(from = 0, to = 500, length = 100)
SSE <- sapply(beta, f1)



beta.SSE = data.frame(beta = beta, SSE = SSE)
plot_ly(data = beta.SSE, x = ~beta, y = ~SSE,
             type = "scatter", mode = 'line')
```

The left panel of Figure 8.6 depicts the SSE versus β for the SIR model. Next, we plug in the estimated $\hat{\beta}$, which is a minimizer of the SSE, to the process. The observed path and the fitted curve for $I(t)$ based on the least-squares method using the grid-searching approach can be found in the left panel of Figure 8.7.

```
beta.hat1 <- beta[which.min(SSE)]
coef(R.ili.data) <- c(Beta = beta.hat1, gamma = 1, N = 14000000,
                      S_0 = 50000, I_0 = 450)
x1 <- trajectory(R.ili.data, format = "data.frame")
sir.grid.fit.df <- left_join(as.data.frame(R.ili.data), x1,
                      by = 'week')



 plot_ly(data = sir.grid.fit.df) %>%
   add_trace(x = ~week, y = ~ilitotal, type = 'scatter',
             mode = 'line', line = list(), name = 'data') %>%
   add_trace(x = ~week, y = ~I, type = 'scatter',
             mode = 'line', line = list(dash = 'dash', color = ""),
             name = 'fitted')
```

However, this grid-search approach will not perform well when we have more than one parameter. In that case, we can use the `optim()` function instead.

```
# General-purpose optimization
f2 <- function(par) {
  params <- c(Beta = par[1], gamma = par[2], N = 14000000,
              S_0 = 50000, I_0 = par[3])
  sse(params)
}
fit2 <- optim(fn = f2, par = c(320, 1, 450))
```

```
beta.hat2 <- fit2$par[1]
gamma.hat2 <- fit2$par[2]
I_0.hat2 <- fit2$par[3]
coef(R.ili.data) <- c(Beta = beta.hat2, gamma = gamma.hat2,
                      N = 14000000, S_0 = 50000, I_0 = I_0.hat2)
x2 <- trajectory(R.ili.data, format = "data.frame")
sir.optim.fit.df <- left_join(as.data.frame(R.ili.data), x2,
                              by = 'week')
```

The observed path and the fitted curve for $I(t)$ based on the least-squares method using general-purpose optimization (`optim()`) can be found in the middle panel of Figure 8.7.

```
plot_ly(data = sir.optim.fit.df) %>%
  add_trace(x = ~week, y = ~ilitotal, type = 'scatter',
            mode = 'line', line = list(), name = 'data') %>%
  add_trace(x = ~week, y = ~I, type = 'scatter',
            mode = 'line', line = list(dash = 'dash'),
            name = 'fitted')
```

8.6.1.2 Maximum Likelihood Approach

Another way to fit the model is using likelihood. Note that the model can be simplified a little bit by defining $b = \beta/N$.

```
R.ili.data2 <- pomp(
  data = subset(ili.data, region == "New England", select = - region),
```

```
                times = "week", t0 = 0,
  skeleton = vectorfield(
    Csnippet("
      double incidence;
      incidence = b * S * I;
      DS = - incidence;
      DI = incidence - gamma * I;")),
  rinit=Csnippet("
      S = S_0;
      I = I_0;"),
  paramnames = c("b", "gamma", "S_0", "I_0"),
  statenames = c("S", "I"))

loglik.normal <- function(params) {
  x <- trajectory(R.ili.data2, params = params)
  sum(dnorm(x = obs(R.ili.data2), mean = x@states["I", ],
            sd = params["sigma"], log = TRUE))
}

f3 <- function(b) {
  params <- c(S_0 = 50000, I_0 = 450, gamma = 1,
              b = b, sigma = 1)
  loglik.normal(params)
}

# (Original fitted beta0)/(previous N) should be close to a solution
b <- seq(from = 200 / 14000000, to = 400 / 14000000, length = 100)
ll <- sapply(b, f3)



b.ll = data.frame(b = b, ll = ll)
plot_ly(data = b.ll, x = ~b, y = ~ll,
                type = "scatter", mode = 'line') %>%
  layout(yaxis = list(title = "log(L)"))
```

The right panel of Figure 8.6 depicts the log-likelihood function versus b for the SIR model. Next, we plug in the estimated b, which is a maximizer of the log-likelihood function, to the process. The observed path and the fitted curve for $I(t)$ based on the maximum likelihood method can be found in the right panel of Figure 8.7.

```
b.hat <- b[which.max(ll)]
coef(R.ili.data2) <- c(S_0 = 50000, I_0 = 450, gamma = 1,
                       b = b.hat, sigma = 1)
x <- trajectory(R.ili.data2, format = "data.frame")
sir.mle.fit.df <- left_join(as.data.frame(R.ili.data2),
                            x, by = 'week')
```

```
plot_ly(data = sir.mle.fit.df) %>%
  add_trace(x = ~week, y = ~ilitotal, type = 'scatter',
            mode = 'line', line = list(),  name = 'data') %>%
  add_trace(x = ~week, y = ~I, type = 'scatter',
            mode = 'line', line = list(dash = 'dash'),
            name = 'fitted')
```

Besides a Gaussian distribution, it is also possible to choose other random components, for example, a Poisson distribution.

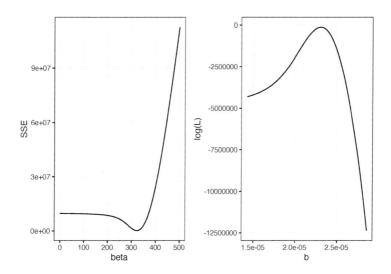

FIGURE 8.6: Left: SSE values versus beta for the least-squares method. Right: log-likelihood function versus b for the maximum likelihood method.

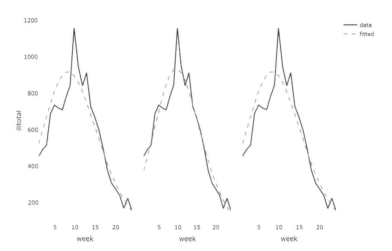

FIGURE 8.7: The observed path and the fitted curve for I(t) for the SIR model. Left: least-squares method using a grid-searching approach. Middle: least-squares method using a general-purpose optimization. Right: maximum likelihood method.

8.6.2   An Application to COVID-19 Data

In the COVID-19 case, the situation becomes slightly more complicated due to the unsatisfying data quality in terms of active cases and recovered cases. Therefore, we modify the algorithm to find the parameters that minimize the distance between the curve of fitted active cases (I) plus removed compartment R and the cumulative positive COVID-19 cases time series. If we use the data observed since the beginning of the pandemic, it is reasonable to set the initial value I_0 to be one and focus on the estimation of other parameters. However, if we choose a different training period, for example, the most recent 60 days, it is possible that the initial value of active cases I_0 may be unobserved or calculated inaccurately. Therefore, the initial value can also be treated as a parameter and estimated based on collected data. In this case, an optimizing function such as `optim()` might be preferred over the grid search. Here is an example of implementing the SIR model to fit the spread of COVID-19 in Los Angeles, California.

```
data("I.county")
data("pop.county")
# The row storing the data for Los Angeles
i <- which(I.county$County == 'Orange' &
           I.county$State == 'California')
```

```
N0 <- pop.county %>%
  filter(County == 'Orange County' & State == 'California') %>%
  pull(population)
est.h <- 60
pred.h <- 12
date.start <- as.Date('2020-10-01')
# Training period
dates.train <- date.start + 0:(est.h - 1)
# Testing period
dates.test <- date.start + est.h + 0:(pred.h - 1)
dates.all <- date.start + 0:(est.h + pred.h - 1)
P.train <- data.frame(Date = 0:(est.h - 1),
                      P_cases = as.numeric(I.county[i,
                      paste0('X', as.character(dates.train,
                            format = '%Y.%m.%d'))]))
P.all <- data.frame(Date = 0:(est.h + pred.h - 1),
                    P_cases = as.numeric(I.county[i,
                    paste0('X', as.character(dates.all,
                          format = '%Y.%m.%d'))]))
```

Now we define a pomp object encoding the data and the model and a loss function sse() that evaluates the loss function value based on the simulated trajectories with given parameters.

```
process.tmp <- pomp(
  data = P.train,
  times = "Date", t0 = 0,
  skeleton = vectorfield(
    Csnippet("
              DS = - Beta*S*I/N;
              DI = Beta*S*I/N - gamma*I;
              DR = gamma*I;")),
  rinit = Csnippet("
                  S = S_0;
                  I = I_0;
                  R = N - S_0 - I_0;"),
  statenames = c("S", "I", "R"),
  paramnames = c("Beta", "gamma", "N", "S_0", "I_0"))


sse <- function(params) {
  x <- trajectory(process.tmp, params = params)
```

```
  discrep <- x@states["I", ] + x@states["R", ] - obs(process.tmp)
  sum(discrep^2)
}
```

Next, we can find the parameters that minimize the SSE using the `optim()` function.

```
# Set initial value of the process
S0 <- N0 - P.train$P_cases[1]
# Assume that the removal rate is known
R.rate <- 0.05
f1 <- function(par) {
  params <- c(Beta = par[2], gamma = R.rate, N = N0,
              S_0 = S0, I_0 = par[1])
  sse(params)
}
# Initial value for parameters
beta0 <- 0.07
I0 <- 1000

fit1 <- optim(fn = f1, par = c(I0, beta0))

I0.hat1 <- fit1$par[1]
beta.hat1 <- fit1$par[2]
process.tmp1 <- process.tmp
pomp::coef(process.tmp1) <- c(Beta = beta.hat1, gamma = R.rate,
                              N = N0, S_0 = S0, I_0 = I0.hat1)
x1 <- trajectory(process.tmp1, times = 0:(est.h + pred.h - 1),
                 format = "data.frame")

valid.df <- data.frame(Date = dates.all,
                       P_cases = P.all$P_cases,
                       SIR_P = x1$I + x1$R,
                       type = c(rep('fitted', est.h),
                                rep('predicted', pred.h)))
```

The observed path, the fitted and predicted curve for $I(t)$ based on the least-squares method using general-purpose optimization for the SIR model are given in Figure 8.8.

```
# Visualization of fitting
plot_ly(data = valid.df) %>%
  add_trace(x = ~Date, y = ~P_cases, type = 'scatter',
            mode = 'line', line = list(dash = 'dash'),
            name = 'infected cases') %>%
  add_trace(x = ~Date, y = ~SIR_P, symbol = ~type,
            mode = 'lines+markers')
```

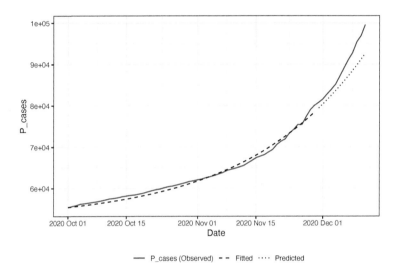

FIGURE 8.8: The observed path, the fitted and predicted curve for $I(t)$ based on the least-squares method using general-purpose optimization for the SIR model in the example of COVID-19.

8.7 Basic and Effective Reproduction Number

8.7.1 Basic Reproduction Number

The **basic reproduction number** (R_0) is a key concept in infectious disease studies, and provides valuable insight into the potential spread of disease. It is defined as the number of secondary cases expected to occur on average from a single (typical) infection in a completely susceptible population.

If $R_0 < 1$, each existing infection results in the formation of fewer new infections. The infection will spread slowly, and the disease will eventually die out.

If $R_0 = 1$, each existing infection results in the emergence of a new infection. The disease will remain alive and stable, but there will not be an outbreak or an epidemic. If $R_0 > 1$, each existing infection multiplies the number of new infections. The infection will spread exponentially, and an outbreak or epidemic may occur.

R_0 is affected by three main factors:

- **Infectious period**: the time frame in which a virus can be transmitted from an infected individual to a susceptible host.

- **Mode of transmission**: the route by which an organism is transferred from one host to another.

- **Contact rate**: the rate at which susceptible individuals meet infected individuals.

At the early stage of an SIR outbreak, when $S(t) \approx N$, the number of infected individuals $I(t)$ is approximated by

$$I(t) \approx I_0 \exp\{(\beta - \gamma - \mu)t\} = I_0 \exp\{(R_0 - 1)(\gamma + \mu)t\},$$

where I_0 is the number of infectious people at time 0, γ^{-1} is the infectious period, and μ^{-1} is the host lifespan. Then, $R_0 \approx \beta/(\gamma + \mu)$ for the SIR model, and an epidemic occurs if the number of infected individuals increases,

$$\beta I(t)\frac{S(t)}{N} - (\mu + \gamma)I(t) > 0 \Rightarrow \beta - (\mu + \gamma) > 0 \Rightarrow R_0 > 1.$$

For the SEIR model, when $S(t) \approx N$, $R_0 = \sigma/(\sigma + \mu) \times \beta/(\gamma + \mu)$.

Studying the dynamics of a newly emerged and rapidly growing infectious disease outbreak, such as COVID-19, is challenging due to the limited amount of data available at the early stage. Initial estimates of the early dynamics of the COVID-19 outbreak in Wuhan, China, suggested a basic reproductive number (R_0) of 2.2–2.7. According to Sanche et al. (2020), the median R_0 for COVID-19 is about 5.7, which is about double an earlier R_0 estimate of 2.2 to 2.7. In this case, 5.7 means that one person infected with COVID-19 can potentially spread the coronavirus to 5 to 6 people, rather than the 2 to 3 people previously thought by researchers. Based on the R_0 of 5.7, in order to stop COVID-19 transmission through vaccination and herd immunity, at least 82% of the population must be immune to it.

Due to an immune or new variant such as Omicron for COVID 19, a population will not be totally susceptible to COVID-19, and the actual reproduction number will be different (typically smaller) than R_0 at the early stage of COVID-19 as time goes by.

8.7.2 Effective Reproduction Number

Due to an infectious individual in a population where some individuals will not be susceptible anymore, the number of susceptible individuals in a population changes over the course of disease. In this section, we introduce the **effective reproduction number** (R_t), which represents the average number of secondary cases per infectious case at a given time t. In infectious disease studies, R_t is one of the most commonly used measures of transmissibility; see, for example, Wallinga and Teunis (2004) and Bettencourt and Ribeiro (2008). Cori et al. (2013) introduced a method to estimate time-varying reproduction numbers $R(t)$ during epidemics, and the method was implemented in the "EpiEstim" R software package. A group of researchers from the London School of Hygiene and Tropical Medicine developed the "EpiNow" R package, which is a substantial extension of the "EpiEstim" package; see Abbott et al. (2020) for a detailed discussion of the approach.

In the following, we show how to estimate R_t using the COVID-19 data as an example. We will use the `estimate_R()` function in the "EpiEstim" package to estimate the effective reproduction number, R_t. To run the following analyses in R, let's load the required packages first.

```
library(IDDA); library(EpiEstim); library(dplyr); library(tidyverse)
```

We consider the cumulative infected COVID-19 cases from the `IDDA:I.state` dataset. Note that the date has the format `X2020.XX.XX`. Thus, we should remove the first `X` using the `substring()` function and make it a "DATE" object using the `as.Date()` function. From this, we can make the daily incidence data from January 22 to March 31, 2020, using the `diff()` function as follows.

```
State.I <- IDDA::I.state %>%
                select(X2020.01.21:X2020.03.31)
seq.date <- as.Date(substring(names(State.I)[-1], 2), format =
    "%Y.%m.%d")
I <- diff(as.numeric(colSums(State.I)))
covid19.incidence <- data.frame(dates = seq.date, I = I)
```

Then, we use the `estimate_R()` function to estimate R_t based on the above data. Note that for the `estimate_R()` function, we need (i) an `incidence` object (with date and incidence), (ii) the time window over which to estimate R_t, and (iii) the distribution of the **serial interval (SI)**, which is the period between the start of symptoms in the primary infector and the onset of symptoms in the infectee. The discrete gamma distribution of the SI is parameterized

by mean and standard deviation. The details can be found from the help
description of discr_si().

While the mean and standard deviation need to be estimated using the paired
date of symptoms onset between infectors and infectees, those data are not
commonly nor easily accessible. In the following example, we use them for
COVID-19 from Rai et al. (2021), which proposed the pooled estimate for the
SI for the outbreak of COVID-19 via meta-analysis. Instead of directly using
the pooled estimates, we take a weighted average of the mean and variance
of the studies in Table 1 of Rai et al. (2021) by considering the sample size.
Then, we have the mean and standard deviation of the SI, 5.52 and 5.44,
respectively. Applying the discr_si() function with the specified mean and
standard deviation, we obtain the discrete gamma distribution as follows.

```
discr_si(k = 0:10, mu = 5.5147, sigma = 5.4353)
```

```
##  [1] 0.00000 0.17179 0.19033 0.12559 0.09443 0.07395
##  [7] 0.05916 0.04795 0.03922 0.03228 0.02671
```

The estimation of R_t can be implemented using estimate_R() with method =
"parametric_si", with the parameters specified above, as follows:

```
res_parametric_si <-
  estimate_R(covid19.incidence, method = "parametric_si",
             config = make_config(list(mean_si = 5.52, std_si = 5.44)))
```

The component R of an object of class estimate_R contains the posterior mean,
standard deviation, and quantiles of R_t for each time window.

```
head(res_parametric_si$R)[,1:5]
```

```
##    t_start t_end Mean(R) Std(R) Quantile.0.025(R)
## 1        2     8   2.073 0.9271            0.6731
## 2        3     9   2.195 0.8959            0.8054
## 3        4    10   1.898 0.7750            0.6967
## 4        5    11   1.687 0.6887            0.6191
## 5        6    12   1.818 0.6873            0.7311
## 6        7    13   1.646 0.6221            0.6617
```

The default time window is a week, and it can be changed using the arguments
t_start and t_end for config. For example,

```
Total.time <- nrow(covid19.incidence)
time.start <- seq(2, Total.time - 13)
time.end <- time.start + 13

res_parametric_si <- estimate_R(covid19.incidence,
                                method = "parametric_si",
                                config = make_config(list(
                                t_start = time.start, t_end = time.end,
                                   mean_si = 5.52, std_si = 5.44)))

head(res_parametric_si$R)[,1:5]
```

```
##   t_start t_end Mean(R) Std(R) Quantile.0.025(R)
## 1       2    15  1.6850 0.4864            0.8707
## 2       3    16  1.5314 0.4421            0.7913
## 3       4    17  1.3002 0.3920            0.6491
## 4       5    18  1.1408 0.3608            0.5471
## 5       6    19  0.9105 0.3219            0.3931
## 6       7    20  1.0664 0.3555            0.4876
```

The plot() function of the output from the estimate_R() function provides
three plots: (i) the time series plot of incidence, (ii) the estimated effective
reproduction number (posterior mean) and 95% credible interval, and (iii) the
discrete gamma distribution of the serial interval with the mean and standard
deviation. The resulting plot is as in Figure 8.9.

```
plot(res_parametric_si, legend = FALSE)
```

When estimating R_t, the distribution of the SI needs to be specified, and
poorly specified information may affect the estimation result. To reduce such
issues, one possible strategy is to integrate the results from several distributions of the SI; for example, the mean and standard deviation of the SI can be
sampled from a truncated normal distributions with additional parameters.

Based on Table 1 of Rai et al. (2021) again, we can take lower and upper
thresholds for a truncated normal distribution for the mean as minimum and
maximum of the mean of the SI, 3.9 and 7.5, and those for the standard
deviation as minimum and maximum of the standard deviation of the SI,

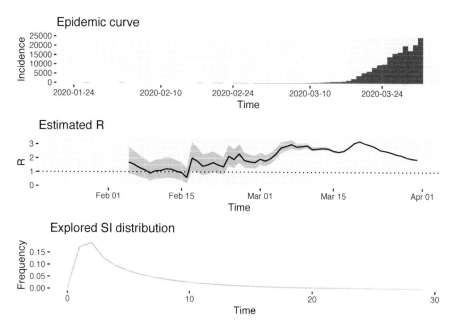

FIGURE 8.9: Top: time series of incidence. Middle: the estimated effective reproduction number (posterior mean) and 95% credible interval based on the `parametric_si` method and 14-day windows. Bottom: the discrete gamma distribution of the serial interval with the mean 5.52 and sd 5.44, respectively.

2.97 and 19.62, among the studies in the meta-analysis of the SI of COVID-19. The standard deviation for a truncated normal distribution for the mean and the standard deviation of the SI are based on those of the mean and standard deviation across studies, 1.16 and 4.52, respectively; the mean and standard deviation of the SI are sampled from a normal(5.52, 1.16) distribution truncated at 3.9 and 7.5, and a normal(5.44, 4.52) distribution truncated at 2.97 and 19.62, respectively. The estimation of R_t can be implemented using `estimate_R()` with `method = "uncertain_si"` to account for uncertainty of the SI distribution, with parameters specified above, as follows:

```
res_uncertain_si <- estimate_R(covid19.incidence,
                          method = "uncertain_si",
                          config = make_config(list(
                          t_start = time.start, t_end = time.end,
                            mean_si = 5.52,
                            min_mean_si = 3.9, max_mean_si = 7.5,
                            std_mean_si = 1.16,
                            std_si = 5.44,
```

```
                           min_std_si = 2.97, max_std_si = 19.62,
                             std_std_si =   4.52)))
```

```
head(res_uncertain_si$R)[,1:5]
```

```
##    t_start t_end Mean(R) Std(R) Quantile.0.025(R)
## 1       2    15  1.6365 0.5218            0.8076
## 2       3    16  1.4853 0.4631            0.7448
## 3       4    17  1.2762 0.4004            0.6239
## 4       5    18  1.1438 0.3647            0.5489
## 5       6    19  0.9282 0.3298            0.4021
## 6       7    20  1.1281 0.3864            0.5104
```

```
plot(res_uncertain_si, legend = FALSE)
```

The lack of information on exact symptom onset between the infector and infectee can lead to a bias of R_t because the serial interval is not based on the incidence of infections, but rather the incidence of symptoms. However, in reality, there must be delayed reports and different incubation periods for each person, even though R_t is a population-averaged value, with sampling bias. Thus, the estimate of R_t should be carefully interpreted by considering such uncertainty; for example, it does not illustrate the current transmission dynamics but depicts the delayed transmission dynamics, especially due to time lag, including the delayed reports and incubation period.

8.7.3 Herd Immunity

It is necessary to lower the basic reproduction number R_0 below one in order to prevent a disease from becoming endemic. Immunization can sometimes help with this.

If a fraction p of the population's newborn members are successfully immunized, the effect is to replace N with $N(1-p)$, bringing the basic reproduction number down to $R_0(1-p)$. The requirement $R_0(1-p) < 1$ gives $1-p < 1/R_0$, or

$$p > 1 - \frac{1}{R_0}.$$

A population is said to have **herd immunity** if a large enough percentage of its members have been immunized to prevent the disease from becoming endemic.

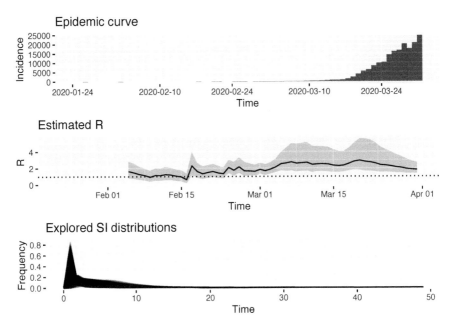

FIGURE 8.10: Top: time series of incidence. Middle: the estimated effective reproduction number (posterior mean) and 95% credible interval based on the `uncertain_si` method and 14-day windows. Bottom: all the serial interval distributions with the mean and sd sampled from truncated normal distributions with specified parameters.

The only diseases for which this has been achieved globally are smallpox (R_0 is approximately 5) and rinderpest, a cattle disease, for which 80% immunization does provide herd immunity.

In the US, epidemiological data show that R_0 for rural populations ranges from 5.4 to 6.3, implying that 81.5% to 84.1% of the population must be vaccinated. In urban areas, R_0 ranges from 8.3 to 13.0, implying that 88.0% to 92.3% of the population must be vaccinated.

8.8   Further Reading

A good introduction to compartmental models in epidemiology can also be found in the following:

- http://en.wikipedia.org/wiki/Compartmental_models_in_epidemiology;

- Held, L., Hens, N., D O'Neill, P., and Wallinga, J. (Eds.). (2019). *Handbook of infectious disease data analysis*. CRC Press.

The following webpages provide many valuable resources with regard to the "deSolve" R package:

- https://cran.r-project.org/web/packages/deSolve/index.html;
- https://desolve.r-forge.r-project.org/;
- https://archives.aidanfindlater.com/blog/2010/04/20/the-basic-sir-model-in-r/.

For further information on the "pomp" package used in this chapter, readers can consult the following web resources.

- https://kingaa.github.io/pomp/;
- https://cran.r-project.org/web/packages/pomp/index.html;
- https://kingaa.github.io/sbied/.

For sources on how to estimate the effective reproduction number in R, readers are referred to the following documents.

- https://cran.r-project.org/web/packages/EpiEstim/;
- https://cran.r-project.org/web/packages/EpiNow2/;
- https://github.com/epiforecasts/EpiNow.

The following article gives a more detailed interpretation of the basic and effective reproduction number:

- Lim, J. S., Cho, S. I., Ryu, S. and Pak, S. I. (2020). Interpretation of the basic and effective reproduction number. *Journal of Preventive Medicine and Public Health*, 53(6), 405.

8.9 Exercises

1. Using data for the outbreak of measles in Niamey in http://kingaa.github.io/clim-dis/parest/niamey.csv, estimate the parameters for the SIR model.

2. We have estimated the parameters by minimizing the SSE between the model-predicted number of cases and the observed data $(\text{prediction} - \text{data})^2$. What would happen if we would like to minimize the squared error on the log scale, i.e., $\{\log(\text{prediction}) - \log(\text{data})\}^2$? What would happen if we would like to minimize the square-root scale, i.e., $(\sqrt{\text{prediction}} - \sqrt{\text{data}})^2$? Try to fit the model with different definitions of the loss function using the niamey data example. What's the "correct" scale to choose?

3. Using the method argument, change the optimization algorithm used by the `optim` function. Examine the impact on your parameter estimations. Try to use other optimizers, such as "nlm," "nlminb," "constrOptim," or the "nloptr" package, etc.

4. Assume instead that the errors are log-normal and have a constant variance. Under what definition of SSE will least-squares and maximum likelihood parameter estimates be the same?

5. Simulate and visualize the dynamics using SIR models with births and deaths. Consider `time = seq(0, 20, by = 1/52)` in years, `N = 100000`, `t_0 = 0`, `S_0 = 100000/12`, `I_0 = 100` and two sets of parameters: (a) $\mu = 1/50$, $\gamma = 365/13$, $\beta = 400$ and (b) $\mu = 1/50$, $\gamma = 365/5$, $\beta = 1000$.

6. According to the model's assumptions, the average host lifespan is $1/\mu$. Explore the differential equations for lifespans of 20 and 200 years to see how the host lifespan affects the dynamics. Consider `time = seq(0, 20, by = 1/52)`, `N = 100000`, `t_0 = 0`, `S_0 = 100000/12`, `I_0 = 100` and parameters: $\gamma = 365/13$, $\beta = 400$.

7. Compare and contrast the dynamics of the SIR and SEIR models for the parameters $\mu = 1/50$, $\gamma = 365/13$, $\beta = 400$ and assuming that for example, $E(0) = 100$, the latent period has a duration of 8 days, and immunity lasts for 10 years in the SEIR model.

9

Time Series Analysis of Infectious Disease Data

Predicting the severity and speed of transmission of the disease is crucial to resource management and developing strategies to control the spread of infectious diseases. Forecasting goals can also be classified as long-term or short-term forecasts. Long-term forecasts can assist predict peak or severity, whilst short-term forecasts can be used by local agencies to direct resource allocation or hospitals to predict case burden in the coming week; see Altieri et al. (2021), Wang et al. (2021b) and Kim et al. (2021).

This chapter focuses on the short-term predictions of the number of infected cases or deaths. Using various examples in COVID-19 prediction, we present how to conduct predictive analysis in a given area via time series analysis, a well-known method in statistics for prediction. This chapter introduces basic knowledge of time series analysis, descriptive techniques, and two main popular methods for forecasting: exponential smoothing and autoregressive integrated moving average models.

9.1 Datasets and R Packages

9.1.1 Data

The dataset `state.long` from the IDDA package contains the cumulative number of cases reported daily for different states in the US mainland. In this chapter, we will base our analysis on the state-level time series.

```
# devtools::install_github('FIRST-Data-Lab/IDDA')
library(IDDA)
data(state.long)
```

Surveillance data collected in infectious disease studies, such as the cumulative number of cases, are often count data. We treat these count variables as continuous in this chapter. Even though the predicted values will be non-integers (with decimals), we can simply round these values to integers. In fact, as long as the mean of the observations is not close to zero, there is no big problem treating the count data as continuous. The rationale is that count data can be approximated by a normal distribution when the mean (or expected value) is high enough. However, if the mean of the observations is relatively small or close to zero, it is critical to model as count data, and models should take into account that observations are nonnegative integers. In statistics, count data is frequently modeled using a generalized linear model (GLM) with a Poisson or negative binomial distribution. Chapter 10 will discuss employing the Poisson regression to model and forecast the count time series.

9.1.2   R Package "fable"

We will use the R packages "fable," "feasts," "forecast," "tsibble" and "dplyr," which together offer various functions for visualizing and analyzing essential time series components. In addition, we use "ggplot2" and "patchwork" for plotting.

```
library(dplyr); library(fable); library(feasts); library(forecast)
library(tsibble); library(ggplot2); library(patchwork)
```

The R package "fable" contains a set of commonly used univariate and multivariate time series forecasting models, such as exponential smoothing through state space models and autoregressive integrated moving average (ARIMA) models. It provides the tools to evaluate, visualize, and combine models in a workflow consistent with the "tidyverse." Details of how to use "fable" for time series analysis are provided by Hyndman and Athanasopoulos (2018).

9.2   An Introduction to Time Series Analysis

A time series is a set of observations recorded sequentially. Data visualization is usually the initial step in a time series study. Many characteristics of the data, such as patterns, odd findings, changes through time, and correlations between variables, can be displayed using graphs. The features observed in data plots should subsequently be included in the forecasting models as much as possible. The type of data decides the forecasting method to employ, and it also defines which graphs to use. Therefore, before generating graphs, we need to first set up our time series in R.

9.2.1   Tsibble Objects

A time series can be thought of as a list of numbers, along with some information about when those numbers were recorded. This information can be stored as a "tsibble" object in R. Beyond the tibble-like representation, key, comprising single or many variables, is used to uniquely identify observational units throughout time (index). In a proper "tsibble" object, index and key uniquely identify each observation.

Take the state.long data for example,

```
head(state.long)
```

```
## # A tibble: 6 x 7
##    State    Region Division        pop DATE       Infected
##    <chr>    <fct>  <fct>         <int> <date>        <int>
## 1 Alabama South  East South~ 4.89e6 2020-12-31   361226
## 2 Alabama South  East South~ 4.89e6 2020-12-30   356820
## 3 Alabama South  East South~ 4.89e6 2020-12-29   351804
## 4 Alabama South  East South~ 4.89e6 2020-12-28   347894
## 5 Alabama South  East South~ 4.89e6 2020-12-27   345623
## 6 Alabama South  East South~ 4.89e6 2020-12-26   343456
## # ... with 1 more variable: Death <int>
```

We can turn this into a "tsibble" object using the as_tsibble() function:

```
state.ts <- as_tsibble(state.long, key = State)
```

```
## Using `DATE` as index variable.
```

```
head(state.ts)
```

```
## # A tsibble: 6 x 7 [1D]
## # Key:        State [1]
```

```
## State     Region Division       pop DATE        Infected
## <chr>     <fct>  <fct>        <int> <date>          <int>
## 1 Alabama South  East South~ 4.89e6 2020-01-22          0
## 2 Alabama South  East South~ 4.89e6 2020-01-23          0
## 3 Alabama South  East South~ 4.89e6 2020-01-24          0
## 4 Alabama South  East South~ 4.89e6 2020-01-25          0
## 5 Alabama South  East South~ 4.89e6 2020-01-26          0
## 6 Alabama South  East South~ 4.89e6 2020-01-27          0
## # ... with 1 more variable: Death <int>
```

The summary above shows that this is a "tsibble" object, which contains 16,905 rows and seven columns. The object is uniquely identified by the key: State. It informs us that there are separate time series in the "tsibble" for each of the 48 adjoining US states and the District of Columbia.

A "tsibble" allows multiple time series to be stored in a single object. The state.long dataset contains the cumulative number of infected cases and deaths for each mainland state in the US and District of Columbia.

9.2.2   Working with Tsibble Objects

Several functions in the "dplyr" package can be used to work with "tsibble" objects, including mutate() and select(). To illustrate these further and some other useful functions, we will use the "tsibble" object state.ts.

```
state.ts <- as_tsibble(state.long, key = State) %>%
  group_by(State) %>%
  mutate(Infected = Infected/1000) %>%
  mutate(YDA_Infected = lag(Infected, order_by - DATE)) %>%
  mutate(YDA_Death = lag(Death, order_by = DATE)) %>%
  mutate(Y.Infected = Infected - YDA_Infected) %>%
  mutate(Y.Death = Death - YDA_Death) %>%
  dplyr::filter(!is.na(Y.Infected)) %>%
  dplyr::filter(!is.na(Y.Death)) %>%
  dplyr::select(-c(YDA_Infected, YDA_Death))
```

```
## Using `DATE` as index variable.
```

```
head(state.ts)
```

```
## # A tsibble: 6 x 9 [1D]
## # Key:        State [1]
## # Groups:     State [1]
##   State   Region Division       pop DATE       Infected
##   <chr>   <fct>  <fct>        <int> <date>          <dbl>
## 1 Alabama South  East South~ 4.89e6 2020-01-23         0
## 2 Alabama South  East South~ 4.89e6 2020-01-24         0
## 3 Alabama South  East South~ 4.89e6 2020-01-25         0
## 4 Alabama South  East South~ 4.89e6 2020-01-26         0
## 5 Alabama South  East South~ 4.89e6 2020-01-27         0
## 6 Alabama South  East South~ 4.89e6 2020-01-28         0
## # ... with 3 more variables: Death <int>,
## #   Y.Infected <dbl>, Y.Death <int>
```

This "tsibble" uses the mutate() function to create two extra variables: Y.Infected and Y.Death, which are the daily new infected count and the daily new deaths for each state.

We can use the filter() function to extract the data for the state of Florida, for example:

```
Florida.ts <- state.ts %>%
  dplyr::filter(State == "Florida")
```

Next, we simplify the resulting object by selecting five variables that we need in subsequent analysis.

```
Florida.ts <- Florida.ts %>%
  dplyr::select(State, Infected, Death, Y.Infected, Y.Death)
head(Florida.ts)
```

```
## # A tsibble: 6 x 6 [1D]
## # Key:        State [1]
## # Groups:     State [1]
##   State   Infected Death Y.Infected Y.Death DATE
##   <chr>      <dbl> <int>      <dbl>   <int> <date>
## 1 Florida        0     0          0       0 2020-01-23
## 2 Florida        0     0          0       0 2020-01-24
## 3 Florida        0     0          0       0 2020-01-25
## 4 Florida        0     0          0       0 2020-01-26
## 5 Florida        0     0          0       0 2020-01-27
## 6 Florida        0     0          0       0 2020-01-28
```

Note that the index variable DATE would be returned even if it was not explicitly selected as it is required for a "tsibble."

9.2.3   Drawing Time Series Plots

To further examine the data, we now use the autoplot() to draw some time series plots; see the top left panel of Figure 9.1.

```
Florida.ts %>% autoplot(Y.Death)
```

We can also use the ggplot() to draw a time series plot. The top right panel of Figure 9.1 shows the time series plot created using the ggplot().

```
ggplot(Florida.ts, aes(x = DATE, y = Y.Death))
```

We can draw multiple time series on the same plot using the "ggplot2" package. The time series plot shown in the bottom panel of Figure 9.1 illustrates the daily new infected cases and deaths in Florida.

```
ggplot(Florida.ts, aes(x = DATE)) +
    geom_line(aes(y = Y.Infected, color = "Y.Infected (thousand)")) +
    geom_line(aes(y = Y.Death, color = "Y.Death")) +
    ylab("Count") + xlab("Day") +
    guides(color = guide_legend(title = "Forecasts")) +
    theme(legend.position = "bottom")
```

We can use the ggplot() function to draw the time series plot of the daily new deaths for each of the midwest states.

```
state.ts %>%
    dplyr::filter(Region == "Midwest") %>%
    ggplot(aes(x = DATE, y = Y.Death,
               group = State, color = State)) +
    geom_line() +
    ylab("Number of deaths") + xlab("Day")
```

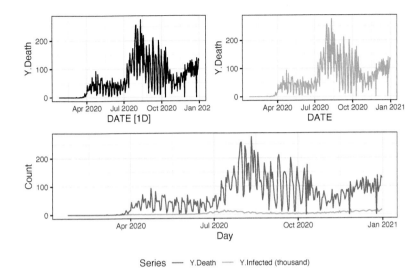

FIGURE 9.1: Top left: the time series plot of the daily new deaths in Florida using the `autoplot()` function. Top right: the time series plot of the daily new deaths in Florida using the `ggplot()` function. Bottom: the time series plot of the daily new infected and death counts in Florida.

The `autoplot()` function is another useful function to draw multiple time series on one plot. Below we show how to use `autoplot()` to draw the time series plot of the daily new deaths for each of the Midwest states.

```
state.ts %>%
    dplyr::filter(Region == "Midwest") %>%
    autoplot(Y.Death) +
    ylab("Number of deaths") + xlab("Day")
```

These two methods generate the same figure, which is shown in Figure 9.2.

After making the time series plot, we look for the following features on the plot.

- **Trend:** long-term upward or downward movement that might be extrapolated into the future.

- **Periodicity:** repetition in a regular pattern (usually peaks and troughs).

- **Seasonality:** periodic behavior of a known period (i.e., a periodic surge in disease incidence corresponding to seasons or other calendar periods).

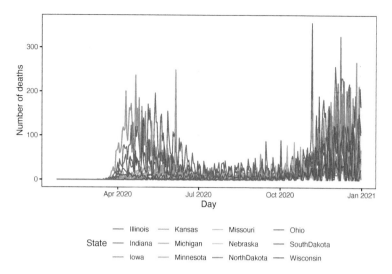

FIGURE 9.2: The time series plot of the daily new deaths for each of the midwest states. Top: using the `ggplot()` function. Bottom: using the `auto-plot()` function.

- **Heteroskedasticity:** changing variance, particularly with changing level.

- **Dependence:** positive (successive observations are similar) or negative (successive observations are dissimilar).

- **Structural break** when a time series abruptly changes at a point in time.

- **Missing data** and **outliers** in a time series.

It is a widespread misconception that cyclic behavior and seasonal behavior are the same things, but they are not. The variations are cyclic if they do not have a defined frequency; the pattern is seasonal if the frequency is constant and linked to some feature of the calendar.

9.2.3.1 Time Series Seasonal Plot

We can produce a time series seasonal plot using the `gg_season()` function in the `feasts` R package, which provides a collection of features, decomposition methods, statistical summaries, and graphics functions to analyze tidy time series data.

A seasonal plot is similar to a regular time series plot, with the exception that the x-axis displays data from each season. This plot format makes the underlying seasonal pattern clear, which is especially important for determining when the pattern changes. The time series plot of daily new deaths for each week is shown in Figure 9.3.

```
library(feasts)
Florida.ts %>% gg_season(Y.Death, period = "week")
```

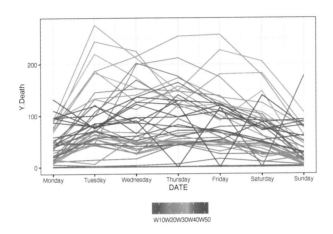

FIGURE 9.3: The time series plot of the daily new deaths for each week.

9.2.3.2 A Lag Plot

We can draw a lag plot to show the time series against lags of itself using the gg_lag() in the R package "feasts."

```
library(feasts)
Florida.ts %>%
   gg_lag(y = Y.Death, geom = "point")
```

The seasonal period (here, a weekly cycle) is often colored to identify how each season (each day in a week) correlates with others. From Figure 9.4, one sees that the relationship is strongly positive at lag 7, reflecting a strong weekly cycle in the data.

9.2.4 Objectives of Time Series Analysis

Typically, a time series analysis is done for the following purposes: (1) To provide an interpretable data model, though it is not always a major emphasis in

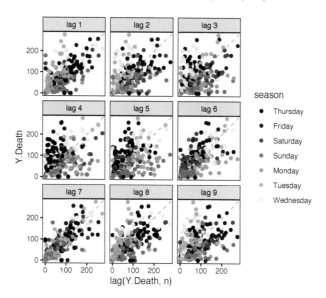

FIGURE 9.4: The lag plot of the daily new deaths for Florida.

time series analysis. It often involves multivariate series and allows the testing of scientific hypotheses. (2) To predict future series values, which is a very common application of time series analysis. (3) To provide a compact description of data since a good predictive model can be used for data compression.

To achieve these objectives, we often take a probabilistic approach. In this case, observations can be considered as realizations of random variables. In time series, random variables are typically not identically distributed or independent. They may have different means due to trend, seasonality, and different variances. The dependence among the random variables may be positive or negative. To make things easier, we can eliminate heteroskedasticity via transformation (e.g., log), trend and seasonality, and then model the remainder as dependent but identically distributed series.

9.2.5   Stationarity

It is hard to model dependence in time series analysis if it changes with time. It is often much easier to model dependence in the stationary case. Roughly speaking, stationary means **probabilistic properties** of series do not change with time.

There are two versions of stationarity. Before introducing them, we first define the mean and covariance function for a time series. Let $\{X_t\}$ be a time series with $\mathrm{E}(X_t^2) < \infty$.

The **mean function** of $\{X_t\}$ is defined as

$$\mu_X(t) = \mathrm{E}(X_t).$$

The **covariance function** of $\{X_t\}$ is defined as

$$\gamma_X(s,t) = \mathrm{Cov}(X_t, X_s) = \mathrm{E}[(X_s - \mu_X(s))(X_t - \mu_X(t))]$$

for all integers s and t.

A time series $\{X_t\}$ is **strictly stationary** if its joint probability distributions do not change with time. That is, for any positive integer k and integers $t_1, \ldots, t_k$ and h,

$$(X_{t_1}, X_{t_2}, \ldots, X_{t_k}) \overset{d}{=} (X_{t_1+h}, X_{t_2+h}, \ldots, X_{t_k+h})$$

where "$\overset{d}{=}$" denotes equality in probability distribution. If $\{X_t\}$ is independent and identically distributed (IID), then $\{X_t\}$ is strictly stationary. Strict stationarity is a very strong modeling assumption, which is hard to verify in practice, and often stronger than necessary for useful results. If $\{X_t\}$ is a sequence of independent random variables that follow the same normal distribution with mean zero, we call $\{X_t\}$ **IID Gaussian noise**.

A time series $\{X_t\}$ is **weakly stationary** if its mean and covariance structures do not change over time. That is, for all integers t and h: (i) $\mathrm{Var}(X_t) < \infty$; (ii) $\mathrm{E}[X_t]$ does not depend on t; and (iii) $\mathrm{Cov}(X_t, X_{t+h})$ does not depend on t. These three conditions above imply respectively: (i) all means, variances, and covariances exist; (ii) all means are constant (rules out trend, seasonality); and (iii) all variances are constant (rules out heteroskedasticity).

Weakly stationary is also known as covariance stationary, second-order stationary, or just stationary, and it is much weaker than strictly stationary.

9.2.6 Autocovariance and Autocorrelation

The autocovariance and autocorrelation functions are essential tools for time series analysis.

For a weakly stationary time series $\{X_t\}_{t=1}^n$, the **autocovariance function (ACVF)** at lag h is defined as

$$\mathrm{Cov}(X_t, X_{t+h}) = \gamma(h),$$

a function of h only.

The **autocorrelation function** (ACF) of $\{X_t\}$ at lag h is defined by

$$\rho(h) = \frac{\gamma(h)}{\gamma(0)} = \frac{\mathrm{Cov}(X_t, X_{t+h})}{\sqrt{\mathrm{Var}(X_t)\mathrm{Var}(X_{t+h})}}.$$

Next, we describe how to estimate the ACVF and the ACF.

Let $\bar{X}_n = n^{-1}\sum_{t=1}^{n} X_t$ be the sample mean of time series $\{X_t\}_{t=1}^{n}$. The **sample ACVF** at lag h (based on $\{X_t\}_{t=1}^{n}$) is defined as

$$\hat{\gamma}(h) = \frac{1}{n}\sum_{t=1}^{n-|h|} (X_t - \bar{X}_n)(X_{t+|h|} - \bar{X}_n), \ |h| < n,$$

which estimates $\gamma(h) = \mathrm{E}[(X_t - \mathrm{E}X_t)(X_{t+|h|} - \mathrm{E}X_t)]$.

The **sample ACF** at lag h based on $\{X_t\}_{t=1}^{n}$ is defined as

$$\hat{\rho}(h) = \hat{\gamma}(h)/\hat{\gamma}(0), \ |h| < n,$$

which estimates $\rho(h) = \gamma(h)/\gamma(0)$.

Figure 9.5 shows the sample ACF plot of the daily new deaths in Florida.

```
Florida.ts %>%
  ACF(Y.Death) %>%
  autoplot()
```

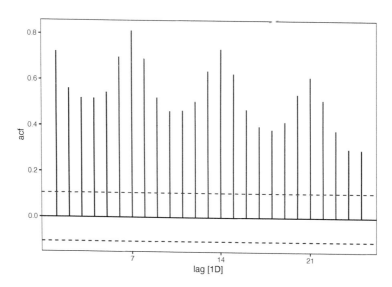

FIGURE 9.5: The ACF plot of the daily new deaths in Florida.

The dashed lines in this graph indicate whether the correlations are significantly different from zero. We see that $\hat{\rho}(7)$ is higher than for the other lags, and this is due to the weekly pattern in the data.

When data has a trend, observations close in time are also close in size. As a result, autocorrelations for small lags tend to be significant and positive. The ACF of a time series with a trend has positive values that gradually decline as the lags increase.

The `gg_tsdisplay()` function in the R package "feasts" provides plots of a time series along with its ACF and a third customizable graphic of either a partial autocorrelation function (PACF), histogram, lagged scatterplot or spectral density. For example, we consider the lag-7 differenced series derived from the daily deaths in Florida. Figure 9.6 displays the time series plot, its ACF plot, and PACF plot.

```
Florida.ts %>%
  gg_tsdisplay(difference(Y.Death, 7),
               plot_type = 'partial', lag = 36) +
  labs(y = "Lag 7 differenced")
```

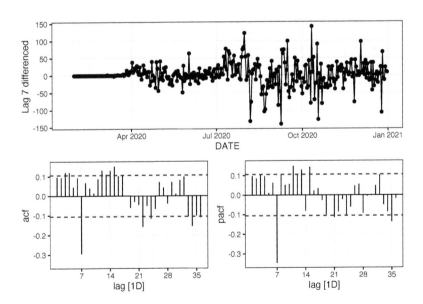

FIGURE 9.6: Time series plot, ACF plot and PACF plot of lag-7 differenced data.

9.3   Time Series Decomposition

Time series data can show a variety of patterns. Therefore, splitting a time series into numerous components indicating different underlying pattern categories is typically useful. A time series is usually decomposed into three components in the literature: a trend component, a seasonal component, and a remainder component. In the following, we consider the following classical decomposition:

$$X_t = m_t + s_t + Y_t, \tag{9.1}$$

where m_t is the **trend** at t (non-random often), s_t is a function with known period d, referred to as **seasonality** (non-random), and Y_t represents the random noise that is often stationary.

In time series analysis, we typically estimate and extract m_t and s_t first so that the residual Y_t will be stationary. Next, we need to find a satisfactory probabilistic model for Y_t, and use it together with m_t and s_t for prediction and simulation of $\{X_t\}$.

9.3.1   Box-Cox Transformations

When decomposing a time series, it can be helpful to transform or adjust the data first to simplify the subsequent decomposition or analysis. Transformations help to stabilize the variance. In the following, we introduce the family of Box-Cox transformations, which is very popular in time series analysis. For time series $\{Y_t\}$, let $\{W_t\}$ be its transformation, which is defined as

$$W_t = \begin{cases} \log(Y_t), & \lambda = 0; \\ (Y_t^\lambda - 1)/\lambda, & \lambda \neq 0. \end{cases}$$

It is worth noting that

- $\lambda = 1$: no substantive transformation;
- $\lambda = 1/2$: square root plus linear transformation;
- $\lambda = 0$: natural logarithm. This is a simple way to force forecasts to be positive;
- $\lambda = -1$: inverse plus 1;
- λ close to zero: $Y_t^\lambda \approx 1$ behaves like logs.

In general, the transformation results are relatively insensitive to the value of λ. It is recommended to choose a simple value of λ to make the explanation easier. If some $Y_t = 0$, then we must set $\lambda > 0$. If some $Y_t < 0$, then no power transformation is possible unless all Y_t are adjusted by adding a constant to all values.

We must reverse the transformation (or back-transform) to obtain forecasts on the original scale. The reverse Box-Cox transformations are given by

$$Y_t = \begin{cases} \exp(W_t), & \lambda = 0; \\ (\lambda W_t + 1)^{1/\lambda} - 1, & \lambda \neq 0. \end{cases}$$

Examples of Box-Cox transformations and reverse transformations are given in Section 9.8.2.

9.3.2 Methods for Estimating the Trend

A trend usually needs to be estimated to fit a time series model. This section describes some popular methods for trend estimation.

9.3.2.1 Polynomial Regression

Polynomial regression is a simple curve fit or approximation model, where the number of cases is approximated locally with polynomials of degree d as shown in the following:

$$Y_t = \beta_0 + \beta_1 t + \beta_2 t^2 + \cdots + \beta_d t^d + Z_t,$$

where $\{Z_t\}$ are IID Gaussian noise.

The top panel of Figure 9.7 is the 14-day-ahead forecast of the daily deaths for Florida using the linear regression method.

```
death_lmfit <- Florida.ts %>%
  model(TSLM(Y.Death ~ DATE))
```

The bottom panel of Figure 9.7 shows the 14-day-ahead forecast of the daily new deaths for Florida using the linear regression method with the seasonal component: season(7).

```
death_lmsfit <- Florida.ts %>%
  model(TSLM(Y.Death   ~ DATE + season(7)))
```

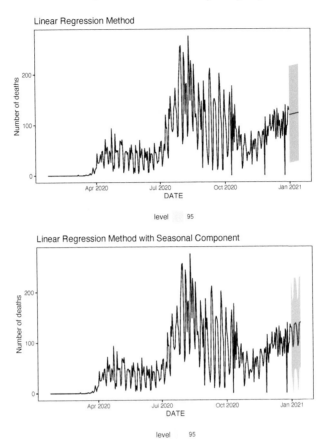

FIGURE 9.7: Two-week-ahead forecast of the daily new deaths for Florida. Top: the linear regression method. Bottom: the linear regression method with the seasonal component.

9.3.2.2   Moving Average Filtering

For an integer $q > 0$ and filtering parameters $\{a_{-q}, a_{-q+1}, \dots, a_0, a_1, \dots, a_q\}$, we consider the following moving average filter:

$$\hat{m}_t = \sum_{k=-q}^{q} a_k X_{t-k}, \ q+1 \leq t \leq n - q.$$

For example, if $a_k = 1/(1+2q)$, then

$$\hat{m}_t = \frac{1}{1+2q} \sum_{k=-q}^{q} m_{t-k} + \frac{1}{1+2q} \sum_{k=-q}^{q} Y_{t-k}$$

is the two-sided moving average. We call this an m-**MA smoothing**, meaning a moving average of order $m = 1+2q$. This method allows linear trend function $m_t = c_0 + c_1 t$ to pass without distortion:

$$\sum_{k=-q}^{q} a_k m_{t-k} = m_t.$$

The m-MA can be easily done using `slide_dbl()` from the "slider" package, which applies a function to "sliding" time windows. We can use the `mean()` function to specify the window size. For a small q, $\hat{m}_t$ is closer to m_t. Intuitively one wants to choose q as small as the data will allow; however, there is always a trade-off between the bias of the estimator and its variance. A smaller q results in less bias but more variance. In contrast, a large q results in more bias but less variance. For example, we consider $q = 2$ and $q = 7$ below to have a 5-MA smoothing and a 15-MA smoothing, respectively.

```
Florida.ts.MA5 <- Florida.ts %>%
  mutate(
    `5-MA` = slider::slide_dbl(Y.Death, mean,
              .before = 2, .after = 2, .complete = TRUE)
  )
Florida.ts.MA15 <- Florida.ts %>%
  mutate(
    `15-MA` = slider::slide_dbl(Y.Death, mean,
              .before = 7, .after = 7, .complete = TRUE)
  )
```

To see what the trend-cycle estimate looks like, we plot the above two moving average trends along with the original data in Figure 9.8.

```
Florida.ts %>% autoplot(Y.Death, color = 'black') +
  autolayer(Florida.ts.MA5, `5-MA`, color = "red", size = 1) +
  autolayer(Florida.ts.MA15, `15-MA`, color = "cyan3", size = 1) +
  labs(y = "Number of deaths", x = "Date",
      title = "Florida daily new deaths with moving average trend")
```

9.3.2.3   Simple Exponential Smoothing

Consider the following non-seasonal model with trend:

$$Y_t = m_t + Z_t,$$

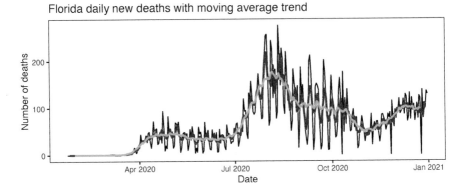

FIGURE 9.8: Florida daily new deaths (black) with the 5-MA (red) and 15-MA (cyan) smoothing of the trend.

where $EZ_t = 0$. We can estimate m_t using simple exponential smoothing. For any fixed parameter α $(0 < \alpha < 1)$, we consider the estimator $\hat{m}_t$ of m_t defined by the following recursions:

$$
\begin{aligned}
\hat{m}_1 &= Y_1, \text{ and} \\
\hat{m}_t &= \alpha Y_t + (1-\alpha)\hat{m}_{t-1} \\
&= \alpha Y_t + (1-\alpha)\{\alpha Y_{t-1} + (1-\alpha)\hat{m}_{t-2}\} \\
&= \alpha Y_t + (1-\alpha)\alpha Y_{t-1} + (1-\alpha)^2 \alpha Y_{t-2} + (1-\alpha)^3 \hat{m}_{t-3} \\
&= \ldots
\end{aligned}
$$

with exponentially decreasing weights on previous observations: $\alpha(1-\alpha)^0$ on Y_t, $\alpha(1-\alpha)^1$ on Y_{t-1}, $\alpha(1-\alpha)^2$ on Y_{t-2}, and so on. In the above exponential smoothing, α is a tuning parameter that controls the performance of the smoothing. Specifically, if α is close to 1, there will be less bias but more variance in the estimation; if α is close to 0, there will be more bias but less variance in the estimation.

This method is particularly suitable for forecasting data with no clear trend or seasonal pattern. The ETS() function returns forecasts and other information for the exponential smoothing forecasts. There are several versions of the ETS models; see Hyndman and Athanasopoulos (2018). We can specify the details of the ETS model by choosing different options in the error, trend and season, and specials.

- The form of the error term: either additive ("A") or multiplicative ("M"). If the error is multiplicative, the data must be non-negative.
- The form of the trend term: either none ("N"), additive ("A"), multiplicative ("M") or damped variants ("Ad", "Md").

- The form of the seasonal term: either none ("N"), additive ("A") or multiplicative ("M").

```
ets_fit <- Florida.ts %>%
  model(ETS(Y.Death
           ~ error("A") + trend("N") + season("N"),
           opt_crit = "mse"))
```

The ETS() also allows us to extend the simple exponential smoothing to make the forecasting of data with a trend and seasonality.

```
etss_fit <- Florida.ts %>%
  model(`ETS` = ETS(Y.Death
                    ~ error("A") + trend("A") + season("A")))
```

The bottom panel of Figure 9.9 displays the observed and fitted daily new deaths using the extended exponential smoothing method with trend and seasonality components.

9.3.3 Seasonal Component

The methods for estimating and eliminating trends discussed above can be naturally adapted to eliminate both trend and seasonality in the general model, as shown below.

Consider the classical decomposition in model (9.1), where s_t has cyclic behavior of known period d, i.e., there is a perfect repetition in s_t: $s_t = s_{t+d}$. We assume that $\sum_{j=0}^{d-1} s_{t-j} = 0$ for model identifiability. Note that if $\sum_{j=0}^{d-1} s_{t-j} = c$, we can always have $\sum_{j=0}^{d-1}(s_{t-j} - c/d) = 0$.

For some time series, we can easily identify the seasonal components using a time series plot, or ACF plot, or smoothing the series. Once we have identified the seasonal component, we can model it by simple differencing (e.g., $X_t - X_{t-12}$) or moving average. We can also consider the dummy variables approach:

$$
s_t =
\begin{cases}
r_1 & t = 1, 1 + d, 1 + 2d, 1 + 3d, ... \\
r_2 & t = 2, 2 + d, 2 + 2d, 2 + 3d, ... \\
\vdots & \\
r_{d-1} & t = d - 1, 2d - 1, 3d - 1, ... \\
-\sum_{k=1}^{d-1} r_k & t = d, 2d, 3d, ...
\end{cases}
$$

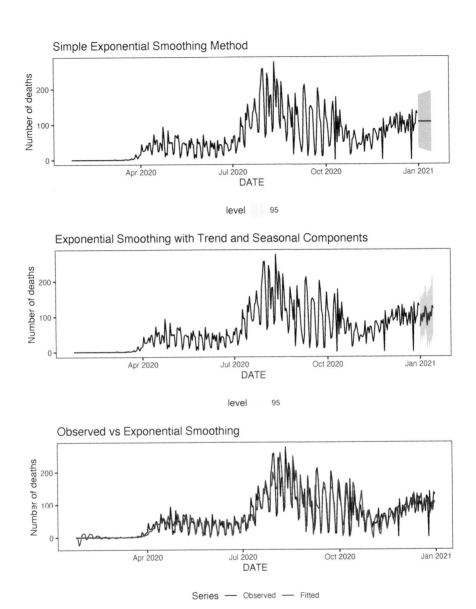

FIGURE 9.9: Top: two-week-ahead forecast of the daily new deaths for Florida using the simple exponential smoothing method. Center: two-week-ahead forecast of the daily new deaths for Florida using the extended exponential smoothing method with trend and seasonality components. Bottom: time series plot of the observed (black) and fitted daily new deaths (red).

Then, we can consider the regression of X_t on dummy variables s_t over t

$$\min \sum_{t=1}^{n} (X_t - s_t)^2$$

for $r_1, \ldots, r_{d-1}$.

9.3.4 Trend and Seasonality Estimation

Let $\{X_t\}$ be a time series. The following describes a typical procedure often used to estimate the trend and seasonal components in a time series analysis.

Step 1. Form a preliminary estimate $\hat{m}_t$ of the trend by passing data through filter/smoothing that eliminates s_t as much as possible.

Step 2. Subtract the trend estimate from the data: $u_t = x_t - \hat{m}_t$.

Step 3. Obtain a seasonal pattern estimate $\{\hat{s}_j : j = 1, \ldots, d\}$.

Step 4. Replicate $\{\hat{s}_j\}$ as needed to form estimate $\{\hat{s}_t\}$ of $\{s_t\}$.

Step 5. Form deseasonalized data: $d_t = x_t - \hat{s}_t$.

Step 6. Use deseasonalized data to get the final estimate $\hat{m}_t$ of the trend.

9.3.4.1 Seasonal and Trend Decomposition Using Loess (STL)

The STL is a versatile and robust approach for decomposing time series created by Cleveland et al. (1990). STL stands for "Seasonal and Trend decomposition Using Loess," and Loess is a method for estimating nonlinear relationships.

The algorithm iteratively updates the trend and seasonal components, removing seasonal values and smoothing the remains to discover the trend. The overall level is added to the trend component after being removed from the seasonal component. This procedure is repeated several times. The residuals from the seasonal plus trend fit make up the final component.

Below we will demonstrate how to use the STL() to decompose the time series of the daily new deaths in Florida.

```
# Time series decomposition
dcmp <- Florida.ts %>%
  model(STL(Y.Death))
components(dcmp) %>% head()
```

```
## # A dable: 6 x 8 [1D]
## # Key:      State, .model [1]
## # :        Y.Death = trend + season_week + remainder
##    State   .model DATE        Y.Death      trend season_week
##    <chr>   <chr>  <date>        <int>      <dbl>       <dbl>
## 1 Florida STL(Y~ 2020-01-23        0    -0.0252     -0.0201
## 2 Florida STL(Y~ 2020-01-24        0    -0.0200      0.143
## 3 Florida STL(Y~ 2020-01-25        0    -0.0149      0.00779
## 4 Florida STL(Y~ 2020-01-26        0    -0.0115     -0.0664
## 5 Florida STL(Y~ 2020-01-27        0    -0.00824     0.0291
## 6 Florida STL(Y~ 2020-01-28        0    -0.00406    -0.0109
## # ... with 2 more variables: remainder <dbl>,
## #   season_adjust <dbl>
```

We will decompose the time series of the daily new deaths in Florida as shown in the top panels of Figure 9.1. Figure 9.10 shows the trend of the time series.

```
Florida.ts %>%
  autoplot(Y.Death, color = "black") +
  autolayer(components(dcmp), trend, color = "red") +
  labs(y = "Number of deaths",
       title = "Daily New Deaths with Trend")
```

Next, we can draw the STL decomposition: trend, days of the week effect and remainder, using the `autoplot()` function.

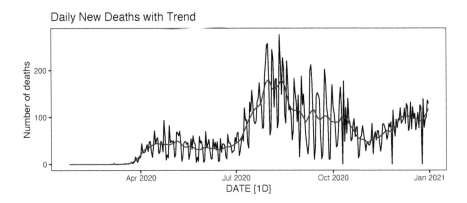

FIGURE 9.10: The observed (black) and the trend component (red) of the daily new deaths time series in Florida.

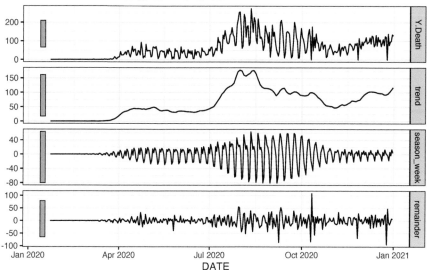

FIGURE 9.11: The trend, seasonality and residuals of the daily new deaths time series in Florida of fitted STL decomposition.

```
components(dcmp) %>% autoplot()
```

The `STL()` function also allows us to choose the trend window `trend(window = )` and the seasonal window `season(window = )`, which controls how rapidly the trend and seasonal components can change. The smaller the value, the more rapid the changes. The trend window is the number of consecutive observations to be used when estimating the trend cycle; the season window is the number of consecutive days to estimate each value in the seasonal component. For example, Figure 9.12 shows the trend, seasonality and residuals of the daily new deaths time series in Florida based on `trend(window = 15)` and `season(window = 7)`. Both trend and seasonal windows should be odd numbers.

```
Florida.ts %>%
  model(STL(Y.Death ~ trend(window = 15) + season(window = 7),
            robust = TRUE)) %>%
  components() %>%
  autoplot()
```

STL decomposition
Y.Death = trend + season_week + remainder

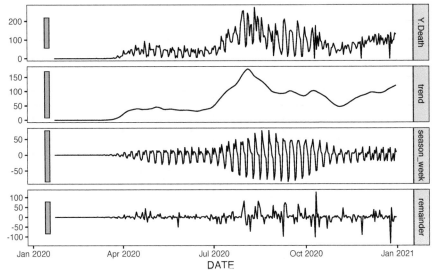

FIGURE 9.12: The trend, seasonality and residuals of the daily new deaths time series in Florida based on `trend(window = 15)` and `season(window = 7)`.

9.4 Simple Time Series Forecasting Approaches

We use data-driven prediction approaches without considering other aspects, such as the disease spread mechanism. We describe each approach in detail in the following subsections.

9.4.1 Average Method

Denote the time series data by $\{Y_t\}$, and consider the following model:

$$Y_t = \mu + Z_t,$$

where $\{Z_t\}$ are IID Gaussian noise.

The average method assumes that forecasts of all future values are equal to the average (or "mean") of the historical data. If we let the historical data be denoted by $Y_1, \ldots, Y_n$, and let $\widehat{Y}_{n+h|n}$ be the estimate of Y_{n+h} based on the historical data, then we can write the forecasts as

$$\hat{Y}_{n+h|n} = \frac{1}{n} \sum_{t=1}^{n} Y_t.$$

For a numeric vector or time series of class ts y, the function `MEAN(y)` returns an i.i.d model applied to y. The `forecast(h = )` returns the forecasts and prediction intervals for Y_{n+h} via the average method, and h is the number of periods for forecasting. We consider the daily new deaths time series in Florida from December 4 to December 17, 2020, as training data. The top left panel of Figure 9.13 shows the two-week-ahead forecast and 95% prediction intervals for the daily new deaths in Florida based on the average method.

```
Florida.ts %>% filter_index("2020-12-04" ~ "2020-12-17") %>%
    model(MEAN(Y.Death))%>%
    forecast(h = 14) %>%
    autoplot(Florida.ts, level = 95, title = "Average Method") +
    labs(y = "Number of deaths", title = "Average Method")
```

9.4.2   Random Walk Forecasts

The random walk model assumes that

$$Y_t = Y_{t-1} + Z_t,$$

where $\{Z_t\}$ are IID Gaussian noise.

The random walk approach simply sets all forecasts to be the value of the last observation. That is,

$$\hat{Y}_{n+h|n} = Y_n.$$

The function `RW(y)` or `NAIVE(y)` together with `forecast(h)` provide the random walk forecasts and prediction intervals for Y_{n+h}.

```
Florida.ts %>% filter_index("2020-12-04" ~ "2020-12-17") %>%
  model(RW(Y.Death))%>%
    # NAIVE(Y.Death) is an equivalent alternative
    forecast(h = 14) %>%
    autoplot(Florida.ts, level = 95) +
    labs(y = "Number of deaths", title = "Random Work Method")
```

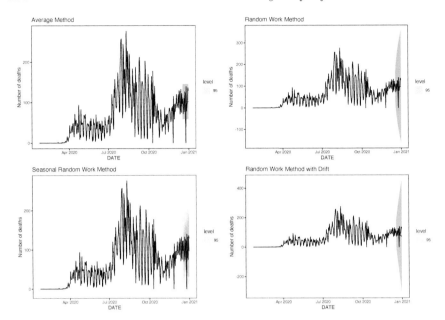

FIGURE 9.13: Two-week-ahead forecast of the daily new deaths for Florida. Top left: average method. Top right: random walk method. Bottom left: seasonal random walk method. Bottom right: random walk with drift method.

9.4.3 Seasonal Random Walk Forecasts

For highly seasonal data, a similar strategy can be used. In this case, we set each forecast equal to the last observed value from the same time of the previous period. Formally, the forecast for time $n + h$ is written as

$$\hat{Y}_{n+h|n} = Y_{n+h-m(k+1)},$$

where m is the seasonal period, and $k = [(h-1)/m]$, that is, the integer part of $(h-1)/m$. For COVID-19 data, we often observe the seven-day cycle; see Wang et al. (2021a). So, we consider $m = 7$, and k is the number of complete weeks in the forecast period prior to time $n + h$.

The function `SNAIVE(y)` with `forecast(h)` provides the seasonal random walk forecasts and prediction intervals for Y_{n+h}.

```
Florida.ts %>% filter_index("2020-12-04" ~ "2020-12-17") %>%
  model(SNAIVE(Y.Death)) %>%
    forecast(h = 14) %>%
    autoplot(Florida.ts, level = 95) +
    labs(y = "Number of deaths", title = "Seasonal Random Work Method")
```

9.4.4 Random Walk with Drift Method

The random walk with drift model is

$$Y_t = c + Y_{t-1} + Z_t,$$

where $\{Z_t\}$ are IID Gaussian noise.

A variation on the random walk method allows the forecasts to increase or decrease over time, where the amount of change over time (called the **drift**) is set to be the average change seen in the historical data. Forecasts are given by

$$\hat{Y}_{n+h|n} = \hat{c}h + Y_n = Y_n + h\left(\frac{Y_n - Y_1}{n - 1}\right).$$

We use the RW(~ drift()) with forecast(h) to make an h-step-ahead forecast.

```
Florida.ts %>% filter_index("2020-12-04" ~ "2020-12-17") %>%
    model(RW(Y.Death ~ drift())) %>%
    forecast(h = 14) %>%
    autoplot(Florida.ts, level = 95)  +
    labs(y = "Number of deaths",
         title = "Random Work Method with Drift")
```

9.4.5 Displaying All Forecasting Results

Now, let us display all the forecasting results based on the previous methods together. Figure 9.14 shows the comparison among different methods.

```
# Fit the models using 4 different methods
death_fit <- Florida.ts %>%
  filter_index("2020-12-04" ~ "2020-12-17") %>%
  model(
    Mean = MEAN(Y.Death),
    `RW` = RW(Y.Death),
    `Seasonal naïve` = SNAIVE(Y.Death),
    `RW-Drift` = RW(Y.Death ~ drift())
  )
# Generate forecasts for the next 2 weeks
death_fc <- death_fit %>% forecast(h = 14)
# Show the forecasts in one plot
death_fc %>%
  #Show the point forecasts only without prediction intervals
```

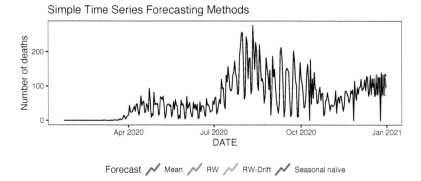

FIGURE 9.14: Two-week-ahead forecast of the daily new deaths for Florida using four different methods.

```
autoplot(Florida.ts, level = NULL, size = 1) +
labs(y = "Number of deaths",
     title = "Simple Time Series Forecasting Methods") +
guides(color = guide_legend(title = "Forecast"),
       linetype = guide_legend(title = "Forecast"))
```

9.4.6   Distributional Forecasts and Prediction Intervals

9.4.6.1   Forecast Distributions

For any $h = 1, \ldots, H$, let Y_{n+h} be the $(n+h)$th observation. Next, let $\hat{Y}_{n+h|n}$ be its forecast based on data up to time n. A forecast $\hat{Y}_{n+h|n}$ is (usually) the mean of the conditional distribution $Y_{n+h} \mid Y_1, \ldots, Y_n$. Most time series models produce normally distributed forecasts. The forecast distribution describes the probability of observing any future value.

Under the models introduced in Sections 9.4.1 to 9.4.4, if residuals are normal and uncorrelated with standard deviation $\hat{\sigma}$, then the h-step-ahead forecast follows a normal distribution. The following describes the distribution of $\hat{Y}_{n+h|n}$ based on each of the abovementioned models.

- **Mean:** $\hat{Y}_{n+h|n} \sim N(\bar{Y}, (1 + 1/n)\hat{\sigma}^2)$.
- **Random walk:** $\hat{Y}_{n+h|n} \sim N(Y_n, h\hat{\sigma}^2)$.
- **Seasonal random walk:** $\hat{Y}_{n+h|n} \sim N(Y_{n+h-m(k+1)}, (k+1)\hat{\sigma}^2)$, where k is the integer part of $(h-1)/m$.

- **Random walk with drift:** $\hat{Y}_{n+h|n} \sim N\{Y_n + (n-1)^{-1}h(Y_n - Y_1), n^{-1}h(n + h)\hat{\sigma}^2\}$.

Note that when $h = 1$ and n is large, these all give the same approximate forecast variance: $\hat{\sigma}^2$.

9.4.6.2 Prediction Intervals

A **prediction interval** gives a region within which we expect Y_{n+h} to lie with a specified probability. Assuming forecast errors are normally distributed, then a 95% prediction interval is

$$\hat{Y}_{n+h|n} \pm 1.96\hat{\sigma}_h,$$

where $\hat{\sigma}_h$ is the standard deviation of the h-step distribution. When $h = 1$, $\hat{\sigma}_h$ can be estimated from the residuals.

We can use the `hilo()` function in the "fable" R package to convert the forecast distributions into intervals. By default, 80% and 95% prediction intervals are returned. We can use the `level` argument to control coverage. The function `unpack_hilo()` allows a hilo() column to be unpacked into its component columns: "lower," "upper," and "level."

```
fc_result <- Florida.ts %>%
  filter_index("2020-12-04" ~ "2020-12-17") %>%
  model(RW(Y.Death ~ drift())) %>%
  forecast(h = 14) %>%
  hilo(level = 95)

unpack_hilo(fc_result, `95%`) %>% head()
```

```
## # A tsibble: 6 x 7 [1D]
## # Key:        State, .model [1]
##    State    .model              DATE         Y.Death .mean
##    <chr>    <chr>               <date>        <dist> <dbl>
## 1 Florida RW(Y.Death ~ d~ 2020-12-18  N(95, 1340)  95.2
## 2 Florida RW(Y.Death ~ d~ 2020-12-19  N(93, 2886)  93.5
## 3 Florida RW(Y.Death ~ d~ 2020-12-20  N(92, 4639)  91.7
## 4 Florida RW(Y.Death ~ d~ 2020-12-21  N(90, 6597)  89.9
## 5 Florida RW(Y.Death ~ d~ 2020-12-22  N(88, 8762)  88.2
## 6 Florida RW(Y.Death ~ d~ 2020-12-23 N(86, 11133)  86.4
## # ... with 2 more variables: 95%_lower <dbl>,
## #   95%_upper <dbl>
```

9.5 Residual Diagnostics and Accuracy Evaluation

9.5.1 Residual Diagnostics

The fitted values and residuals from a model can be obtained using the augment() function. In the above example, we saved the fitted models as death_lmfit. So we can simply apply the augment() function to this object to compute the fitted values and residuals for all models.

```
augment(death_lmfit) %>% head()
```

```
## # A tsibble: 6 x 7 [1D]
## # Key:        State, .model [1]
##    State .model DATE       Y.Death .fitted .resid .innov
##    <chr> <chr>  <date>       <int>   <dbl>  <dbl>  <dbl>
## 1 Flor~ TSLM(~ 2020-01-23       0    7.05  -7.05  -7.05
## 2 Flor~ TSLM(~ 2020-01-24       0    7.37  -7.37  -7.37
## 3 Flor~ TSLM(~ 2020-01-25       0    7.70  -7.70  -7.70
## 4 Flor~ TSLM(~ 2020-01-26       0    8.02  -8.02  -8.02
## 5 Flor~ TSLM(~ 2020-01-27       0    8.35  -8.35  -8.35
## 6 Flor~ TSLM(~ 2020-01-28       0    8.68  -8.68  -8.68
```

Residuals are useful for determining if a model has captured all of the information from the data. The model can probably be improved if patterns can be seen in the residuals. Residues having the following attributes will be produced by a good forecasting method:

- There is no correlation between the residuals. If there are correlations between residuals, then the residuals contain information that should be employed in forecasting.
- The mean of the residuals is zero. The forecasts are biased if the residuals have a mean other than zero.
- The variance of the residuals is constant.
- The residuals follow a normal distribution.

Any forecasting method that does not satisfy these properties can be improved.

The residuals obtained from forecasting this series using the linear regression method with seasonal components are shown in Figure 9.15.

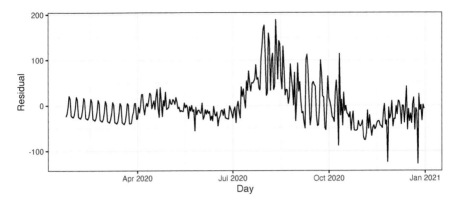

FIGURE 9.15: Residual plot based on the linear regression method with seasonal components.

```
augment(death_lmsfit) %>%
  autoplot(.resid) +
  labs(x = "Day", y = "Residual")
```

The `gg_tsresiduals()` function, which produces a time plot, ACF plot, and histogram of the residuals, is a handy shortcut for creating these residual diagnostic graphs.

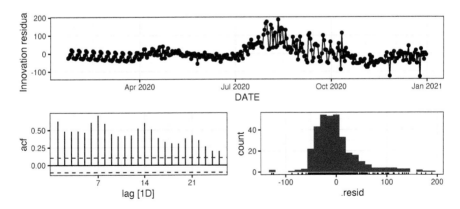

FIGURE 9.16: Time plot, ACF plot and histogram of the residuals based on the linear regression method with seasonal components.

```
gg_tsresiduals(death_lmsfit)
```

The residual plots obtained from forecasting this series using the extended
exponential smoothing method with trend and seasonality components are
shown in Figure 9.17.

```
gg_tsresiduals(etss_fit)
```

These graphs demonstrate that the ETS technique generates projections that
appear to take into consideration all available information. The residuals are
mean zero with significant correlation. Except for one outlier, the variation of
the residuals remains relatively consistent across the historical data, as seen
by the time plot of the residuals. As a result, the residual variance can be considered constant. This can also be observed in the histogram of the residuals.
According to the histogram, the residuals appear to be normal. Consequently,
forecasts from this method are likely to be highly accurate, and prediction
intervals calculated using a normal distribution seem reasonable.

Instead of checking whether each sample autocorrelation falls inside the
bounds defined in Figure 9.17, it is also possible to consider a portmanteau
test such as the Ljung–Box test. The hypothesis of IID data is then rejected
at level α if the p-value is smaller than α.

```
augment(etss_fit) %>%
features(.resid, ljung_box, lag = 10, dof = 0)
```

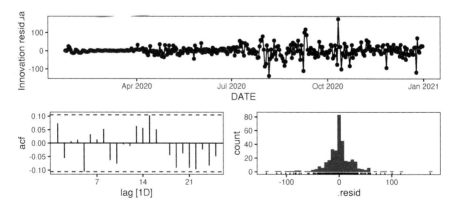

FIGURE 9.17: Time plot, ACF plot and histogram of the residuals based
on the extended ETS method with the trend and seasonal components.

```
## # A tibble: 1 x 4
##    State    .model lb_stat lb_pvalue
##    <chr>    <chr>    <dbl>     <dbl>
## 1 Florida ETS        11.8     0.297
```

9.5.2 Forecasting Accuracy

Cross-validation is a popular technique for model comparison. First, we split the dataset into a subset called the **training set**, and another subset called the **test set**. Then, we train the model on the training set, and record the forecast error on the test set.

Forecast error is defined as the difference between an observed value and its forecast:

$$e_{n+h} = Y_{n+h} - \hat{Y}_{n+h|n},$$

where the training data is given by $\{Y_1, \dots, Y_n\}$.

For a continuous or discrete time series, a point forecast is usually preferred, such as the number of daily confirmed cases. Several measures have been developed to assess the validity of these point forecasts:

$$\text{Mean Absolute Error (MAE)} = \frac{1}{n}\sum |e_t|,$$

$$\text{Mean Squared Error (MSE)} = \frac{1}{n}\sum e_t^2,$$

$$\text{Root Mean Squared Error (RMSE)} = \sqrt{\frac{1}{n}\sum e_t^2},$$

$$\text{Mean Absolute Percentage Error (MAPE)} = \frac{1}{n}\sum |100 e_t/Y_t|,$$

$$\text{Mean Absolute Scaled Error (MASE)} = \frac{1}{n}\sum (|e_t|/Q),$$

where $Q = (n-1)^{-1}\sum_{t=2}^{n} |Y_t - Y_{t-1}|$.

Note that MAE, MSE, RMSE are all scale-dependent. MAPE and MASE are scale-independent, but MAPE is more sensible than MASE. MAPE can be infinite or undefined if $Y_t = 0$ for any t in the period of interest, and have extreme values if any Y_t is close to zero.

In the following, we split the data `Florida.ts` into two parts: a training set (November 20 to December 17, 2020) and a validation set or test set (December 18 ~ December 31, 2020). The model is fit on the training set, and the fitted model is used to predict the responses for the observations in the test set.

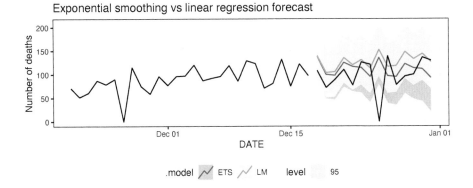

FIGURE 9.18: Two-week-ahead forecast of the daily new deaths in Florida.

Figure 9.18 shows the forecasts of the daily new deaths for December 18 ~ December 31 based on the training data.

```
library(tsibble)
# Set training data from NOV 20 to DEC 17, 2020
train <- Florida.ts %>%
    filter_index("2020-11-20" ~ "2020-12-17")

Reg_fit <- train %>%
    model(`LM` = TSLM(Y.Death   ~ DATE + season(7)),
          `ETS` = ETS(Y.Death
                       ~ error("A") + trend("A") + season("A")))
Reg_fc <- Reg_fit %>% forecast(h = 14)

Reg_fc %>%
    autoplot(train, level = 95) +
    autolayer(filter_index(dplyr::select(Florida.ts, Y.Death),
                    "2020-12-18" ~ .), color = "black") +
    labs(y = "Number of deaths",
        title = "Exponential smoothing vs linear regression forecast")
```

The accuracy() function can be used to obtain the accuracy of the forecast, which is able to automatically extract the relevant periods from the data (Florida.ts in this example) to match the forecasts when computing the various accuracy measures.

```
accuracy(Reg_fc, Florida.ts)
```

```
## # A tibble: 2 x 11
##    .model State    .type    ME  RMSE   MAE   MPE  MAPE
##    <chr>  <chr>    <chr> <dbl> <dbl> <dbl> <dbl> <dbl>
## 1 ETS    Florida Test  -12.3  44.4  32.4  -Inf   Inf
## 2 LM     Florida Test  -28.6  49.1  33.5  -Inf   Inf
## # ... with 3 more variables: MASE <dbl>, RMSSE <dbl>,
## #   ACF1 <dbl>
```

Chapter 5 from Hyndman and Athanasopoulos (2018) provides more details on evaluating models and the corresponding forecasting accuracy.

9.5.3   Selection of the Time Series

How far do we want to look into the past? Using a longer time series doesn't necessarily lead to a more accurate forecast. Since time series modeling aims to generate valuable forecasts, the series should contain data of the same nature as the forecasts required. So the series should not extend so far into the past to include periods during which a different case definition was applied or any other factors resulting in a mean number of cases per reporting interval (a day in our examples) that differs from the mean of recent reporting intervals. On the other hand, the required sample size also depends on the number of parameters to be estimated in the model and the noise level of the data. We may use more historical observations to fit a time series model when there are more model parameters to be estimated, and the amount of noise is larger.

9.6   ARIMA Models

9.6.1   Differencing

When working with time series lags, it is useful to introduce the backward shift operator. Let B be the backward shift operator such that $B^k f(t) = f(t - k)$.

For example,

$$
\begin{aligned}
B^1 t &= t - 1, \\
B^2 t &= t - 2, \\
B^2 t^2 &= (t - 2)^2, \\
Bc &= c \quad (c \text{ is a constant}).
\end{aligned}
$$

Next, we define $1 - B$ as the **difference operator**. Using the difference operator, it is easier to describe the process of differencing, e.g.,

$$(1 - B)X_t = X_t - BX_t = X_t - X_{t-1}.$$

The **second-order difference** is denoted $(1 - B)^2$, which is not the same as a second difference $1 - B^2$. In general, a d**th-order difference** can be written as

$$(1 - B)^d X_t.$$

Here are two examples of differencing.

Example 9.1. If $m_t = \alpha + \beta t$, then the first-order difference of m_t is $(1 - B)m_t = \beta$.

Example 9.2. If $m_t = \alpha + \beta t + \gamma t^2$, then the second-order differences of m_t is $(1 - B)^2 m_t = 2\gamma$.

When combining differences, the backshift notation is particularly handy with the operator being treated by standard algebraic methods. Terms involving B can be multiplied together in particular.

Let $1 - B^k$ be the lag-k difference operator, that is,

$$(1 - B^k)X_t = X_t - X_{t-k}.$$

Applying lag-d differencing to remove seasonality, we have

$$(1 - B^d)s_t = s_t - B^d s_t = s_t - s_{t-d} = 0.$$

In fact, we can show that

$$(1 - B^d) = \underbrace{(1 - B)}_{\text{difference operator}} \underbrace{[1 + B + B^2 + \cdots + B^{d-1}]}_{\text{seasonal summation filter}}$$

noting that $(1 + B + B^2 + \cdots + B^{d-1})s_t = s_t + s_{t-1} + \cdots + s_{t-(d-1)} = \sum_{j=0}^{d-1} s_{t-j} = 0.$

We can use the difference() function in the "tsibble" package, and two important arguments are:

- lag: a positive integer indicating which lag to use;
- differences: a positive integer indicating the order of the difference.

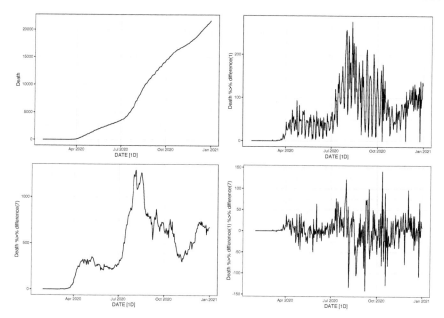

FIGURE 9.19: Time series plots of the death count in Florida. Top left: cumulative death count. Top right: daily new deaths. Bottom left: weekly new deaths. Bottom right: weekly change in daily new deaths.

Figure 9.19 demonstrates various ways to make the difference for the time series of the cumulative deaths in Florida. For example, difference(1) provides the daily new deaths, and difference(7) provides the weekly new deaths.

```
Florida.ts %>%
  autoplot(Death)
Florida.ts %>%
  autoplot(Death %>% difference(1))
Florida.ts %>%
  autoplot(Death %>% difference(7))
Florida.ts %>%
  autoplot(Death %>% difference(1) %>% difference(7))
```

The following provides another method to eliminate the trend and seasonality using differencing.

Step 1. Apply a lag-d differencing operator to X_t:

$$
\begin{aligned}
(1 - B^d)X_t &= m_t - m_{t-d} + s_t - s_{t-d} + Y_t - Y_{t-d} \\
&= m_t - m_{t-d} + Y_t - Y_{t-d}.
\end{aligned}
$$

Then, the resulting model has a trend component defined by $m_t - m_{t-d}$ and a stochastic component given by $Y_t - Y_{t-d}$.

Step 2. The trend component can be eliminated by applying an appropriate power of differencing operator, say $(1 - B)^{d'}$. Thus,

$$\underbrace{(1 - B)^{d'}}_{} \; \underbrace{(1 - B^d)}_{} X_t = \underbrace{(1 - B)^{d'}}_{} \; \underbrace{(1 - B^d)}_{} m_t + \underbrace{(1 - B)^{d'}}_{} \; \underbrace{(1 - B^d)}_{} Y_t$$

is a model for a series related to $\{X_t\}$ that is free of trend and seasonal components.

9.6.2 ARMA Models

This section introduces an important parametric family of stationary time series, the autoregressive moving-average, or ARMA, processes.

A process $\{X_t\}$ is said to be ARMA(p, q) (for integers $p, q \geq 0$), or an **AutoRegressive (AR) Moving Average (MA)**, with

$$
\begin{aligned}
\text{AR polynomial } \phi(z) &= 1 - \phi_1 z - \phi_2 z^2 - \cdots - \phi_p z^p, \text{ and} \\
\text{MA polynomial } \theta(z) &= 1 + \theta_1 z + \theta_2 z^2 + \cdots + \theta_q z^q,
\end{aligned}
$$

if $\{X_t\}$ satisfies

$$\phi(B) X_t = \theta(B) Z_t \quad \text{for all integers } t,$$

with respect to some $\{Z_t\} \sim \text{WN}(0, \sigma^2)$, where B is the backward shift operator. That is,

$$
\begin{aligned}
\phi(B) X_t &= X_t - \phi_1 X_{t-1} - \phi_2 X_{t-2} - \cdots - \phi_p X_{t-p} \\
&= Z_t + \theta_1 Z_{t-1} + \theta_2 Z_{t-2} + \cdots + \theta_q Z_{t-q} = \theta(B) Z_t.
\end{aligned}
$$

The time series $\{X_t\}$ is said to be an autoregressive process of order p (or AR(p)) if $\theta(z) \equiv 1$, and a moving-average process of order q (or MA(q)) if $\phi(z) \equiv 1$.

Example 9.3. The AR(1) model is a special case of the ARMA models, and we can denote it as ARMA(1,0):

$$X_t = \phi X_{t-1} + Z_t$$

with AR polynomial: $\phi(z) = 1 - \phi z$ and MA polynomial: $\theta(z) = 1$.

Example 9.4. The MA(1) is also a special case of the ARMA models, denoted as ARMA(0,1):

$$X_t = Z_t + \theta Z_{t-1}$$

with AR polynomial: $\phi(z) = 1$ and MA polynomial: $\theta(z) = 1 + \theta z$.

9.6.3 ARIMA Models

We consider a generalization of the ARMA models to incorporate a wide range of nonstationary series.

A process $\{Y_t\}$ is said to be ARIMA(p, d, q) (for integers $p, d, q \geq 0$), or an **AutoRegressive (AR) Integrated (I) Moving Average (MA)**, if $\{Y_t\}$ satisfies

$$(1 - B)^d Y_t = X_t \sim \text{ARMA}(p, q) \text{ for all integers } t.$$

For example, suppose $X_t \sim \text{ARMA}(p, q)$, and let

$$Y_t = Y_{t-1} + X_t = \cdots = Y_0 + \sum_{j=1}^{t} X_j, \ t = 1, 2, \ldots,$$

then $Y_t \sim \text{ARIMA}(p, 1, q)$.

Figure 9.20 illustrates a typical procedure for building an appropriate ARIMA model.

How do we identify the parameters p, d and q in ARIMA(p, d, q)? A time series plot can help identify preliminary values of d. If we conduct too little differencing, then the time series might not be stationary. However, if we conduct too much differencing, we may introduce extra dependence in the time series. To identify p and q, it is often helpful to examine the sample ACF and PACF of $(1 - B)^d X_t$. Table 9.1 below provides some tips in choosing p and q.

9.6.4 Seasonal ARIMA (SARIMA) Model

A SARIMA model is formed by including additional seasonal terms in the ARIMA models we have seen so far.

TABLE 9.1: ACF and PACF for ARMA models

Model	ACF	PACF
AR(p)	decays	zero for h > p
MA(q)	zero for h > q	decays
ARMA(p,q)	decays	decays

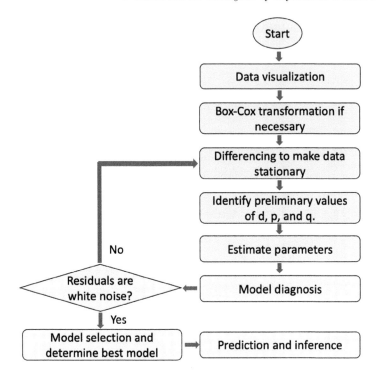

FIGURE 9.20: A procedure to build ARIMA models.

If d and D are nonnegative integers, then $\{X_t\}$ is a **seasonal ARIMA**$(p, d, q) \times (P, D, Q)_s$ **process with period** s if the differenced series

$$Y_t = (1 - B)^d (1 - B^s)^D X_t$$

is a causal ARMA process defined by

$$\phi(B)\Phi(B^s)Y_t = \theta(B)\Theta(B^s)Z_t, \quad Z_t \sim WN(0, \sigma^2),$$

where

$$
\begin{aligned}
\phi(z) &= 1 - \phi_1 z - \phi_2 z^2 - \cdots - \phi_p z^p, \\
\Phi(z) &= 1 - \Phi_1 z - \Phi_2 z^2 - \cdots - \Phi_p z^P, \\
\theta(z) &= 1 + \theta_1 z + \theta_2 z^2 + \cdots + \theta_q z^q, \\
\Theta(z) &= 1 + \Theta_1 z + \Theta_2 z^2 + \cdots + \Theta_q z^Q.
\end{aligned}
$$

A special case of the SARIMA models is the **pure seasonal ARMA model**. For $P, Q \geq 0$ and $s > 0$, we say that a time series $\{X_t\}$ is an ARMA$(P,Q)_s$ process if $\Phi(B^s)X_t = \Theta(B^s)Z_t$, where

$$\Phi(B^s) = 1 - \sum_{j=1}^{P} \Phi_j B^{js},$$

and

$$\Theta(B^s) = 1 + \sum_{j=1}^{Q} \Theta_j B^{js}.$$

Example 9.5. Consider $X_t = Z_t + \Theta_1 Z_{t-12}$, then $\{X_t\}$ is an ARMA$(0,1)_{12}$ process with the following ACF function:

$$\gamma(0) = (1 + \Theta_1^2)\sigma^2,$$
$$\gamma(12) = \Theta_1 \sigma^2,$$
$$\gamma(h) = 0, \text{ for } h = 1, 2, \dots, 11, 13, 14, \dots.$$

Example 9.6. Consider $X_t = \Phi_1 X_{t-12} + Z_t$, then $\{X_t\}$ is an ARMA$(1,0)_{12}$ process with with the following ACF function:

$$\gamma(0) = \frac{\sigma^2}{1 - \Phi_1^2},$$
$$\gamma(12i) = \frac{\sigma^2 \Phi_1^i}{1 - \Phi_1^2},$$
$$\gamma(h) = 0, \text{ for other } h.$$

The ACF and PACF for a seasonal ARMA(P,Q)s are zero for $h \neq si$. For $h = si$, they are analogous to the patterns for ARMA(p,q):

We can estimate an ARIMA model using the ARIMA function, which searches through the model space specified in the specials to identify the best ARIMA model. We can specify an ARIMA model via the formula argument. If the right-hand side of the formula is left blank, the default search space is given by pdq() + PDQ(): that is, a model with candidate seasonal and non-seasonal terms, but no exogenous regressors. To specify a model fully (avoid automatic selection), the intercept and pdq(), PDQ() values need to be given: for example,

```
formula = response ~ 1 + pdq(1, 1, 1) + PDQ(1, 0, 0)
```

In the above, the pdq() option is used to specify non-seasonal components of the model, and the PDQ() option is used to specify seasonal components of the model. To force a non-seasonal fit, specify PDQ(0, 0, 0) in the right-hand side

TABLE 9.2: ACF and PACF for ARMA(P,Q) models

Model	ACF	PACF
AR(P)s	decays	zero for $i > P$
MA(Q)s	zero for $i > Q$	decays
ARMA(P,Q)s	decays	decays

of the model formula. The `period` argument is used in `PDQ()` to specify the periodic nature of the seasonality.

Many criteria have been proposed for the purpose of order determination, such as final prediction error (FPE), Akaike's information criterion (AIC), Bayesian information criterion (BIC), and Akaike's information corrected criterion (AICC) developed by Hurvich and Tsai (1989). We can specify the information criterion used in selecting the model via the `ic` option.

```
ic = c("aicc", "aic", "bic")
```

Then it searches through the model space specified in the specials to identify the best ARIMA model, with the lowest AICc, AIC or BIC value.

9.6.5 Building SARIMA Models

The seasonal lags of the PACF and ACF reveal the seasonal component of an AR or MA model. For `Florida.ts`, after a lag 7 differencing, Figure 9.6 shows the sample ACF and PACF plots. From Figure 9.6, we observe a spike at lag 7 in the ACF and a spike at lag 7 at the PACF but no other significant spikes. Therefore, we can specify `PDQ(1, 1, 1)` in the RHS of the model formula as shown below.

```
sarima111 <- Florida.ts %>%
  model(ARIMA(Y.Death ~ PDQ(1,1,1)))
```

We can use the function `report()` to obtain the formatted model-specific display.

```
  report(sarima111)
```

```
## Series: Y.Death
## Model: ARIMA(1,0,1)(1,1,1)[7]
##
## Coefficients:
##           ar1      ma1     sar1     sma1
##        0.9690  -0.8142   0.2294  -0.7650
## s.e.   0.0173   0.0370   0.1221   0.0961
##
## sigma^2 estimated as 890.9:  log likelihood=-1622
## AIC=3255    AICc=3255    BIC=3274
```

Thus, the fitted SARIMA$(1,0,1)(1,1,1)_7$ can be written as

$$(1 - \phi_1 B)(1 - \Phi_1 B^7)(1 - B^7)X_t = (1 + \theta_1 B)(1 + \Theta_1 B^7)Z_t,$$

where $Z_t \sim WN(0, 890.9)$, and $\phi_1 = 0.9690$, $\theta_1 = -0.8142$, $\Phi_1 = 0.2294$, $\Theta_1 = -0.7650$.

Below, we try to fit more SARIMA models for `Florida.ts`, and conduct the two-week-ahead forecast.

```
death_sarima <- Florida.ts %>%
  model(
    sarima011 = ARIMA(Y.Death ~ PDQ(0,1,1)),
    sarima111 = ARIMA(Y.Death ~ PDQ(1,1,1)),
    stepwise = ARIMA(Y.Death),
    search = ARIMA(Y.Death, stepwise = FALSE)
  )
# Generate forecasts for the next 2 weeks
death_fc <- death_sarima %>% forecast(h = 14)
# Plot forecasts against actual values
death_fc %>%
  autoplot(Florida.ts, level = NULL) +
  labs(y = "Number of deaths", title = "Different ARIMA Forecasts") +
  guides(color = guide_legend(title = "Methods"))
```

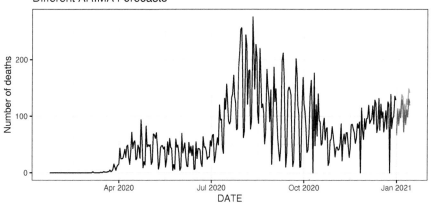

FIGURE 9.21: Two-week-ahead forecast of the daily new deaths for Florida using different ARIMA models.

The `glance()` method returns a one-row summary of each model including descriptions of the model's fit, such as residual variance and information criteria. It is worth noting that the information criteria (AIC, AICc, and BIC) can only be compared between models of the same class and with the same response (after transformations and differencing).

```
death_sarima %>%
  glance() %>%
  arrange(AICc)
```

```
## # A tibble: 4 x 9
##    State   .model    sigma2 log_lik   AIC  AICc   BIC
##    <chr>   <chr>      <dbl>   <dbl> <dbl> <dbl> <dbl>
## 1 Florida search      867.  -1617. 3250. 3251. 3281.
## 2 Florida sarima111   891.  -1622. 3255. 3255. 3274.
## 3 Florida sarima011   897.  -1624. 3255. 3255. 3270.
## 4 Florida stepwise    897.   1624. 3255. 3255. 3270.
## # ... with 2 more variables: ar_roots <list>,
## #   ma_roots <list>
```

9.7 Model Comparison

9.7.1 Exponential Smoothing and ARIMA Models

The two most generally used approaches to time series forecasting are exponential smoothing and ARIMA models, which provide complementary approaches to the problem. The ETS model, as mentioned in Hyndman and Athanasopoulos (2018), describes how unobserved data components (error, trend, and seasonality) vary over time, whereas ARIMA emphasizes data autocorrelations. Furthermore, whereas additive ETS models are special cases of ARIMA models, there are no corresponding ARIMA counterparts for non-additive ETS models. Many ARIMA models, on the other hand, lack exponential smoothing equivalents. Finally, while all ETS models are non-stationary, some ARIMA models are.

Note that we cannot use AIC or AICc to compare ETS and ARIMA because they are in different model classes, and the likelihood is computed in different ways. Instead, we can use the validation comparison by dividing the data into two parts: a training set (November 20 to December 17, 2020) and a validation set or test set (December 18 to December 31, 2020). We can use some common accuracy measures to examine how well the models fit the data.

```
library(urca)
death_fit <- train %>%
  model(
    `ETS` = ETS(Y.Death
                ~ error("A") + trend("A") + season("A")),
    `ARIMA` = ARIMA(Y.Death, stepwise = FALSE)
  )

# Model fitting results of ARIMA
death_fit %>%
  dplyr::select(ARIMA) %>%
  report()
```

```
## Series: Y.Death
## Model: ARIMA(0,1,2)
##
## Coefficients:
##           ma1      ma2
##        -1.344   0.5975
## s.e.    0.231   0.2352
##
## sigma^2 estimated as 605:  log likelihood=-124.8
## AIC=255.6    AICc=256.6   BIC=259.5
```

```
# Evaluate the modeling and forecasting accuracy
death_fit %>%
  accuracy() %>%
  arrange(MASE)
```

```
## # A tibble: 2 x 11
##    State   .model .type      ME  RMSE   MAE   MPE  MAPE
##    <chr>   <chr>  <chr>   <dbl> <dbl> <dbl> <dbl> <dbl>
## 1 Florida ETS    Training -3.99  20.4  14.6  -Inf   Inf
## 2 Florida ARIMA  Training  5.15  23.2  19.1  -Inf   Inf
## # ... with 3 more variables: MASE <dbl>, RMSSE <dbl>,
## #   ACF1 <dbl>
```

It seems that the ETS model outperforms the ARIMA for the series based on the MASE of the training set.

```
# Generate forecasts for the next 2 weeks
death_fc <- death_fit %>% forecast(h = 14)

# Plot forecasts against actual values
death_fc %>%
  autoplot(train, level = 95) +
  autolayer(filter_index(
    dplyr::select(Florida.ts, Y.Death),
    "2020-12-18" ~ .), color = "black") +
  labs(y = "Number of deaths",
       title = "ETS vs ARIMA Forecasts") +
  guides(color = guide_legend(title = "Forecasts"))



# Evaluate the forecasting accuracy based on test set
death_fc %>%
  accuracy(Florida.ts) %>%
  arrange(MASE)



## # A tibble: 2 x 11
##    .model State    .type      ME  RMSE   MAE   MPE  MAPE
##    <chr>  <chr>    <chr>   <dbl> <dbl> <dbl> <dbl> <dbl>
## 1 ARIMA  Florida  Test    -1.95  34.9  25.8  -Inf   Inf
## 2 ETS    Florida  Test   -12.3   44.4  32.4  -Inf   Inf
## # ... with 3 more variables: MASE <dbl>, RMSSE <dbl>,
## #    ACF1 <dbl>
```

Based on the MASE of the test set and Figure 9.22, it seems that the ARIMA model slightly outperforms the ETS for this dataset.

9.7.2 Cross-validation for Time Series Analysis

We split the training set into a training subset and a validation set if any parameters need to be tweaked; see Figure 9.23. On the training subset, the model is trained, and the parameters that minimize error on the validation set are selected. Finally, the model is trained using the chosen parameters on the entire training set, and the error on the test set is recorded.

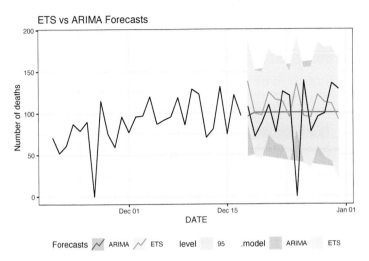

FIGURE 9.22: Two-week-ahead forecast of the daily new deaths for Florida using ETS and ARIMA models.

However, the choice of the test set in Figure 9.23 may seem to be fairly arbitrary, and that choice may mean that our test set error is a poor estimate of the error on an independent test set.

Cross-validation is a popular technique for tuning hyperparameters and producing robust measurements of model performance. Two of the most common types of cross-validation are leave-one-out cross-validation and k-fold cross-validation.

Figure 9.24 illustrates the idea of k-fold cross-validation. First, we randomly split data into k folds, a subset called the training set, and another subset called the test set based on one fold. Focus on the training set, which contains $k - 1$ folds, to train the model and test on the kth fold. Repeat the above k times to get k accuracy measures on 10 different and separate folds. Compute the average of the k accuracies, which is the final reliable number telling us how the model is performing.

Because of the temporal dependencies, traditional cross-validation (such as k-fold) should not be employed when working with time series data.

Cross-validation is more difficult in the case of time series. We can't pick random samples and assign them to the test or training sets because it is pointless to forecast past values using future values. The forecaster must withhold any data regarding events that occur chronologically after the events used for fitting the model to accurately simulate the "real world forecasting environment, in which we stand in the present and forecast the future" (Tashman, 2000).

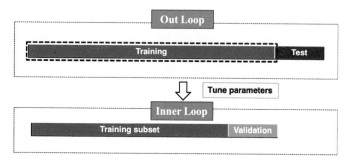

FIGURE 9.23: An illustration of traditional time series validation.

Instead of using k-fold cross-validation, we use hold-out cross-validation for time series data, in which a subset of the data (divided temporally) is reserved for verifying the model performance. The test set data, for example, comes chronologically after the training set, as seen in Figure 9.25. Similarly, the validation subset follows the training subset chronologically. The inner loop operates in the same way as discussed before: the training set is divided into a training subset and a validation set, the model is trained on the training subset, and the parameters that minimize error on the validation set are chosen. The forecast accuracy is derived by averaging over the test sets in the outer loop.

There are three main rolling types which can be used.

- Stretch: extends a growing length window with new data; see Figure 9.25.
- Slide: shifts a fixed length window through the data; see Figure 9.26.
- Tile: moves a fixed length window without overlap; see Figure 9.27.

We can apply the following functions to roll a "tsibble": `stretch_tsibble()`, `slide_tsibble()` or `tile_tsibble()`.

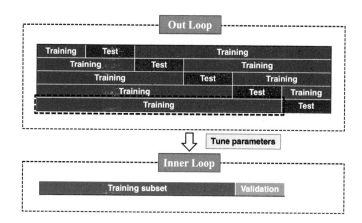

FIGURE 9.24: An illustration of k-fold cross-validation.

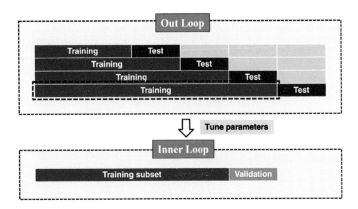

FIGURE 9.25: An illustration of stretch rolling cross-validation for time series.

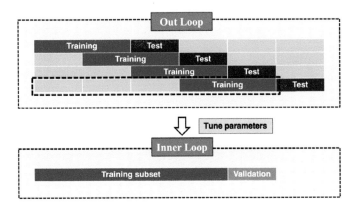

FIGURE 9.26: An illustration of slide rolling cross-validation for time series.

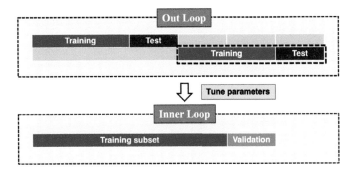

FIGURE 9.27: An illustration of tile rolling cross-validation for time series.

These functions make it quick and easy to roll over a tsibble using observations. They all produce a tsibble with a new column called .id in the key.
With slide_tsibble() and stretch_tsibble(), the output dimension will grow
significantly, and the function is likely to run out of memory if the data is
huge.

For time series cross-validation, stretching window methods are most commonly used.

```
# Split training and test using stretching window
Florida.train <- Florida.ts %>%
  filter_index("2020-04-01" ~ "2020-12-17") %>%
  # Stretch with a minimum length of 60
  # growing by 7 each step
  stretch_tsibble(.init = 60, .step = 7) %>%
  relocate(DATE, State, .id)

head(Florida.train)
```

```
## # A tsibble: 6 x 7 [1D]
## # Key:        State, .id [1]
## # Groups:     State [1]
##    DATE        State     .id Infected Death Y.Infected
##    <date>      <chr>   <int>    <dbl> <int>      <dbl>
## 1 2020-04-01 Florida      1     7.77   100       1.03
## 2 2020-04-02 Florida      1     8.99   144       1.22
## 3 2020-04-03 Florida      1    10.2    169       1.26
## 4 2020-04-04 Florida      1    11.5    194       1.28
## 5 2020-04-05 Florida      1    12.3    220       0.811
## 6 2020-04-06 Florida      1    13.6    253       1.28
## # ... with 1 more variable: Y.Death <int>
```

```
# Training set model fit
FL.fit <- Florida.train %>%
  model(
      `ETS` = ETS(Y.Death
                  ~ error("A") + trend("A") + season("A")),
      `ARIMA` = ARIMA(Y.Death ~ 1 + pdq(1,0,1) + PDQ(0,1,1))
    )
```

```
# Training set accuracy
FL.fit %>% accuracy()
```

```
## # A tibble: 58 x 12
##    State     .id .model .type      ME  RMSE   MAE   MPE
##    <chr>   <int> <chr>  <chr>   <dbl> <dbl> <dbl> <dbl>
##  1 Florida     1 ETS    Train~ -1.67   14.4  11.5 -34.1
##  2 Florida     1 ARIMA  Train~  1.08   15.7  11.3 -26.3
##  3 Florida     2 ETS    Train~ -1.08   14.4  11.2 -27.9
##  4 Florida     2 ARIMA  Train~  1.43   15.3  11.2 -27.8
##  5 Florida     3 ETS    Train~ -0.671  13.9  10.7 -25.6
##  6 Florida     3 ARIMA  Train~  1.41   14.9  10.9 -24.9
##  7 Florida     4 ETS    Train~ -1.00   13.3  10.2 -27.1
##  8 Florida     4 ARIMA  Train~  1.35   14.4  10.3 -23.4
##  9 Florida     5 ETS    Train~ -0.963  13.1   9.85 -24.8
## 10 Florida     5 ARIMA  Train~  1.29   13.9  10.1 -22.9
## # ... with 48 more rows, and 4 more variables:
## #   MAPE <dbl>, MASE <dbl>, RMSSE <dbl>, ACF1 <dbl>
```

```
# 7-day forecast accuracy
period.fc <- 7
FL.fc <- FL.fit %>%
  forecast(h = period.fc) %>%
  group_by(.id) %>%
  mutate(h = row_number()) %>%
  ungroup()

FL.accuracy <- FL.fc %>%
  accuracy(Florida.ts, by = c("h", ".model"))
FL.accuracy$h <- rep(1:period.fc, 2)
```

```
FL.accuracy %>%
  ggplot(aes(x = h, y = RMSE)) +
  geom_line(aes(color = .model)) +
  geom_point(aes(color = .model)) +
  guides(color = guide_legend(title = "Models"))
```

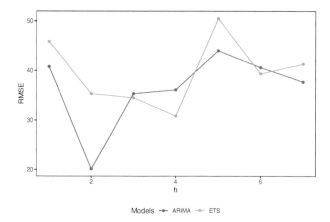

FIGURE 9.28: One week ahead forecast of the daily new deaths for Florida using ETS and ARIMA models.

9.8 Ensuring Forecasts Stay within Limits

It is typical in epidemic forecasting to set projections to be positive, particularly when projecting count time series, or to require them to fall within a certain range ($[a, b]$). Transformations make both of these scenarios reasonably simple to deal with.

9.8.1 Positive Forecasts

To impose a positivity constraint, we can simply work on the log scale. The following is an example using ETS models applied to the daily new deaths time series for Florida.

```
# Obtain the ETS fit with/without log transformation
ETS.fit <- train %>%
  model(
    `ETS` = ETS(Y.Death ~ error("A") + trend("A") + season("A")),
    `logETS` = ETS(log(Y.Death + 1) ~ error("A") +
                   trend("A") + season("A"))
  )


# Generate forecasts for the next 2 weeks
ETS_fc <- ETS.fit %>% forecast(h = 14)
```

```
# Plot forecasts against actual values
ETS_fc %>%
  autoplot(train, level = 95) +
  autolayer(
    filter_index(dplyr::select(Florida.ts, Y.Death),
                 "2020-12-18" ~ .), color = "black") +
  labs(y = "Number of deaths",
       title = "ETS Forecasts with/without log transformation") +
  guides(color = guide_legend(title = "Forecasts"))
```

```
# Residual plot for the ETS without the log transformation
ETS.fit %>%
  select(ETS) %>% gg_tsresiduals(lag = 36)
```

```
# Residual plot for the log transformed EST method
ETS.fit %>%
  select(logETS) %>% gg_tsresiduals(lag = 36)
```

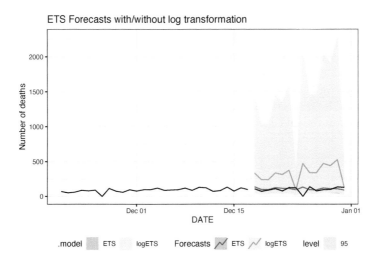

FIGURE 9.29: Two-week-ahead forecast of the daily new deaths for Florida using ETS with/without log transformation.

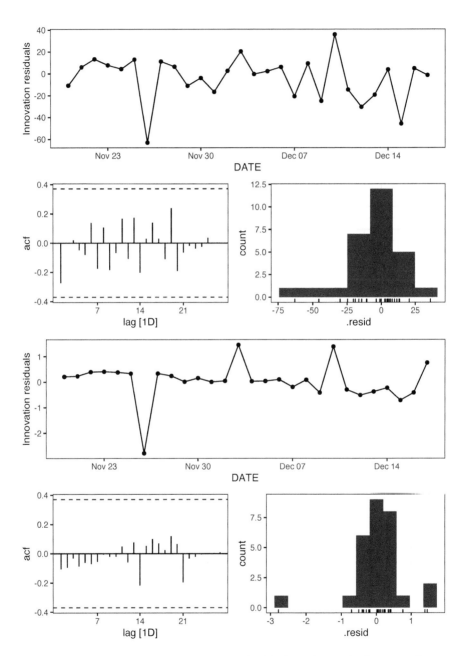

FIGURE 9.30: Top: residual plot for the ETS without the log transformation. Bottom: residual plot for the log transformed EST method.

9.8.2 Forecasts Constrained to an Interval

Sometimes it makes sense to assume that the number of deaths is constrained to lie within $[a, b]$. To handle data constrained to an interval, we can transform the data using a scaled logit transform as follows:

$$y = \log \left(\frac{x - a}{b - x} \right),$$

where x is on the original scale and y is the transformed data. To reverse the transformation, we will use

$$x = \frac{(b - a)e^y}{1 + e^y} + a.$$

This is not a built-in transformation, so we will need to first set up the transformation functions.

```
scaled_logit <- function(x, lower = 0, upper = 1){
  log((x - lower) / (upper - x))
}
inv_scaled_logit <- function(x, lower = 0, upper = 1){
  (upper - lower) * exp(x) / (1 + exp(x)) + lower
}
my_scaled_logit <- new_transformation(scaled_logit,
                                      inv_scaled_logit)
```

Now, we can make the prediction based on the transformed time series. Let us consider the forecast within $[0, 300]$.

```
train %>%
  model(ETS(my_scaled_logit(Y.Death + 1, lower = 0, upper = 300) ~
            error("A") + trend("A") + season("A"))) %>%
  forecast(h = 14) %>%
  autoplot(train, level = 95)
```

9.9 Prediction and Prediction Intervals for Aggregates

We considered the daily new deaths in the above sections, but we may want to forecast the cumulative number of deaths. If the point forecasts are means,

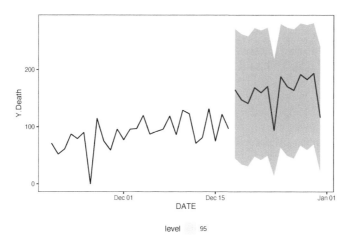

FIGURE 9.31: Two-week-ahead forecast of the daily new deaths for Florida using ETS, constrained to be within [0,300].

then adding them up will give a reasonable estimate of the cumulative count. However, prediction intervals are trickier due to the correlations between forecast errors. A general solution is to use simulations. For example, we consider the forecast of the cumulative death count in the next two weeks.

```
d_ets_fit <-  train %>% model(
    `ETS` = ETS(Y.Death
                ~ error("A") + trend("A") + season("A"))
    )

d.pred.paths <- d_ets_fit %>%
  # Simulate 10000 future sample paths, each of length 14
  generate(times = 10000, h = 14) %>%
  # Sum the results for each sample path
  as_tibble() %>%
  group_by(.rep) %>%
  mutate(.sim = as.integer(.sim)) %>%
  mutate(.sim = replace(.sim, which(.sim < 0), 0)) %>%
  mutate(.csum = cumsum(.sim))
```

We can compute the mean of the simulations, and extract a specified prediction interval at a particular level. For example,

```
d.pred.report <- d.pred.paths %>%
  group_by(DATE) %>%
  summarize(.mean = mean(.csum),
            .lpi95 = quantile(.csum, .025),
            .upi95 = quantile(.csum, .975),
          ) %>%
  mutate(.mean = .mean + tail(train$Death,1),
         .lpi95 = .lpi95 + tail(train$Death,1),
         .upi95 = .upi95 + tail(train$Death,1)
        )
```

Figure 9.32 shows the two-week-ahead forecast of the cumulative number of deaths for Florida using ETS.

```
ggplot(train, aes(DATE, Death)) +
  geom_line() +
  labs(x = "Days", y = "Count",
       title = 'Cumulative deaths and prediction') +
  # Add prediction intervals
  geom_ribbon(mapping = aes(x = DATE,
                            y = .mean,
                            ymin = .lpi95,
                            ymax = .upi95,
                            fill = '95% Prediction intervals'),
              data = d.pred.report, alpha = 0.4) +
  # Add line for predicted values
  geom_line(mapping = aes(x = DATE,
                          y = .mean,
                          color = 'Predicted Value'),
            linetype = "dashed", data = d.pred.report,
            # Set the line type in legend
            key_glyph = "timeseries") +
  scale_color_manual("", values = "red") +
  scale_fill_manual("", values = "pink") +
  guides(color = guide_legend(title = "Series"))
```

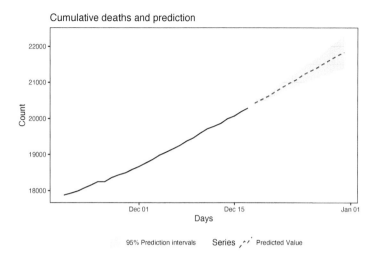

FIGURE 9.32: Two-week-ahead forecast of the cumulative number of deaths for Florida using ETS.

9.10 Outliers and Anomalies

Surveillance systems collect and report data, but their reliability depends on the quality of data and efficient algorithms for detecting ongoing changes in the incidence patterns. In practice, data often contain outlying observations, extreme values, anomalies, and many other messy features. Many data scientists (researchers) suggest that much effort needs to be dedicated to the pre-processing procedure to detect and handle these issues. In this section, we will cover the techniques of dealing with outliers and anomalies.

In a dataset, an **anomaly** is often referred to as a data point that cannot be explained by the base distribution or does not conform to an expected pattern of other observations. On the other hand, an **outlier** is a data point that is distant from other observations, and it usually has a rare chance of occurrence. The two terms are often used interchangeably in practice.

In monitoring the dynamics of infectious diseases, we often encounter three types of anomalies in the time series data: order dependencies violation, point anomalies and subsequence anomalies.

- **Order dependencies violation**: order dependency is widely used in the relational database. For a cumulative count time series $\{Y_t\}$, it can be defined as follows: for any two time points, t_1 and t_2, if $t_1 < t_2$, then $Y_{t_1} \leq Y_{t_2}$.

- **Point anomalies**: an observation that acts abnormally in a given time occurrence when compared to the other values in the time series (global outlier), or to its nearby points (local outlier).

- **Subsequence anomalies**: consecutive observations in time whose joint behavior is unusual, although each observation individually is not necessarily a point anomaly.

Anomalies might be induced in the data for various reasons. For example, the anomalies in COVID-19 could be caused by reasons such as the results released of a large batch of tests, and the change of reporting standard, such as some states starting to report probable cases from a specific date. Anomaly identification is vital because anomalies can hold critical insights into aberrant data behavior. In the following, we focus on the detection of point anomalies or outliers.

Dealing with outliers is one of the earliest data analysis challenges, and since nearly all datasets contain outliers with different percentages, it continues to be one of the most critical problems to solve. Numerous algorithms have been developed to detect and treat outliers in the past decades, especially for univariate datasets.

In the following, we introduce several ways to find outliers using the time series of the daily new deaths for the state of New Jersey; see Figure 9.33. Figure 9.33 shows the time series plot of the daily new death count.

```
# New Jersey daily new deaths
NewJersey.ts <- state.ts %>%
  dplyr::filter(State == "NewJersey") %>%
  dplyr::select(Y.Infected, Y.Death)
head(NewJersey.ts)
```

```
## # A tsibble: 6 x 4 [1D]
## # Key:       State [1]
## # Groups:    State [1]
##    State      Y.Infected Y.Death DATE
##    <chr>           <dbl>   <int> <date>
## 1 NewJersey           0       0 2020-01-23
## 2 NewJersey           0       0 2020-01-24
## 3 NewJersey           0       0 2020-01-25
## 4 NewJersey           0       0 2020-01-26
## 5 NewJersey           0       0 2020-01-27
## 6 NewJersey           0       0 2020-01-28
```

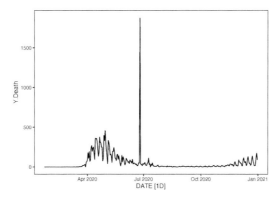

FIGURE 9.33: Time series plot of the daily new death count for New Jersey.

```
# Time series plot death counts in New Jersey
NewJersey.ts %>% autoplot(Y.Death)
```

9.10.1 Empirical Rule

If we can assume the data is approximately normally distributed, we can consider the "empirical rule," which indicates that about 99.7% of the data points lie within three standard deviations below and above the mean of the data. The following code shows the outliers identified under this rule for New Jersey's daily new death count.

```
D_outliers <- NewJersey.ts %>%
  select(DATE, Y.Death) %>%
  filter(
    Y.Death < mean(Y.Death) - 3 * sd(Y.Death) |
    Y.Death > mean(Y.Death) + 3 * sd(Y.Death)
  )
D_outliers
```

```
## # A tsibble: 2 x 3 [1D]
## # Key:       State [1]
## # Groups:    State [1]
##     State      DATE        Y.Death
##     <chr>      <date>        <int>
## 1 NewJersey 2020-04-30        458
## 2 NewJersey 2020-06-25       1877
```

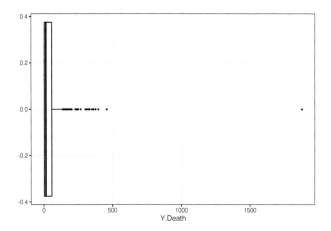

FIGURE 9.34: Boxplots of the daily new death count for New Jersey.

9.10.2 Boxplots

One of the most basic and popular univariate outlier detection techniques is the "Boxplot Rule."

```
NewJersey.ts %>% ggplot(aes(x = Y.Death)) +
  geom_boxplot(outlier.color = "black", outlier.shape = 16,
               outlier.size = 1, notch = FALSE)
```

9.10.3 Outliers in Time Series

In time series, outlier detection is often performed on remainders that removed both the seasonal components and trend components. There are many ways that a time series can be deconstructed to produce remainders, including (1) ARIMA models, (2) machine learning (regression), and (3) seasonal decomposition. In the following, we decompose a time series applying STL() with the argument robust=TRUE to remove trend and/or seasonal components. In the following code, we fit the model for New Jersey's daily new infected count and death count, respectively, using the function STL() without a seasonal component, i.e., season(period = 1), and investigate the outliers based on the remainders. See Figures 9.35 and 9.36 for the STL decomposition.

```
# Fit a non-seasonal STL decomposition: infected series
NJ_I_decomp <- NewJersey.ts %>%
```

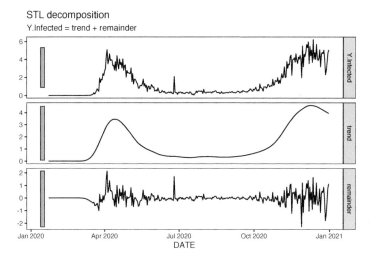

FIGURE 9.35: A non-seasonal STL decomposition for New Jersey's daily new infected count.

```
  model(
    stl = STL(Y.Infected ~ season(period = 1), robust = TRUE)
  ) %>%
  components()
NJ_I_decomp %>% autoplot()
```

```
# Fit a non-seasonal STL decomposition: death series
NJ_D_decomp <- NewJersey.ts %>%
  model(
    stl = STL(Y.Death ~ season(period = 1), robust = TRUE)
  ) %>%
  components()
NJ_D_decomp %>% autoplot()
```

If the outliers are not obvious from the plot, we can perform the "Boxplot Rule" on the remainders for a test.

```
I_outliers <- NJ_I_decomp %>%
  dplyr::filter(
    remainder < quantile(remainder, 0.25) - 3 * IQR(remainder) |
```

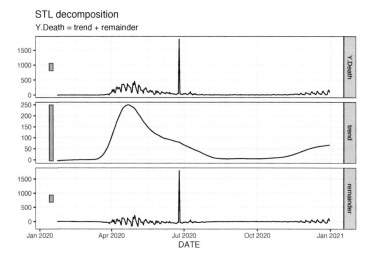

FIGURE 9.36: A non-seasonal STL decomposition for New Jersey's daily new death count.

```
    remainder > quantile(remainder, 0.75) + 3 * IQR(remainder)
  )
I_outliers
```

```
## # A dable: 29 x 7 [1D]
## # Key:      State, .model [1]
## # :        Y.Infected = trend + remainder
##     State     .model DATE       Y.Infected trend remainder
##     <chr>     <chr>  <date>          <dbl> <dbl>     <dbl>
##  1 NewJersey stl    2020-03-25      0.439 1.46      -1.02
##  2 NewJersey stl    2020-04-03      4.37  2.83       1.53
##  3 NewJersey stl    2020-04-04      5.10  2.96       2.14
##  4 NewJersey stl    2020-04-07      4.63  3.24       1.38
##  5 NewJersey stl    2020-04-16      4.43  3.43       0.997
##  6 NewJersey stl    2020-04-23      4.06  3.07       0.993
##  7 NewJersey stl    2020-05-03      3.06  2.23       0.836
##  8 NewJersey stl    2020-06-25      2.11  0.403      1.70
##  9 NewJersey stl    2020-11-10      3.77  2.91       0.868
## 10 NewJersey stl    2020-11-14      4.34  3.30       1.04
## # ... with 19 more rows, and 1 more variable:
## #   season_adjust <dbl>
```

```
D_outliers <- NJ_D_decomp %>%
  dplyr::filter(
    remainder < quantile(remainder, 0.25) - 3 * IQR(remainder) |
    remainder > quantile(remainder, 0.75) + 3 * IQR(remainder)
  )
D_outliers
```

```
## # A dable: 23 x 7 [1D]
## # Key:      State, .model [1]
## # :         Y.Death = trend + remainder
##     State    .model DATE       Y.Death trend remainder
##     <chr>    <chr>  <date>        <int> <dbl>     <dbl>
##  1 NewJersey stl    2020-04-08      270  177.      92.6
##  2 NewJersey stl    2020-04-13       93  218.    -125.
##  3 NewJersey stl    2020-04-14      362  227.     135.
##  4 NewJersey stl    2020-04-15      351  233.     118.
##  5 NewJersey stl    2020-04-16      361  238.     123.
##  6 NewJersey stl    2020-04-17      322  241.      80.8
##  7 NewJersey stl    2020-04-19      134  246.    -112.
##  8 NewJersey stl    2020-04-21      377  249.     128.
##  9 NewJersey stl    2020-04-26       75  243.    -168.
## 10 NewJersey stl    2020-04-27      106  241.    -135.
## # ... with 13 more rows, and 1 more variable:
## #   season_adjust <dbl>
```

9.10.4 Tidy Anomaly Detection for Time Series with "anomalize"

In the examples below, a point anomaly refers to the situation where there is one day of an abrupt increase in the cumulative or daily new time series. There are a few R packages that can be applied to detect point anomalies or outlier for time series. For example, we can use the "anomalize" R package developed by Dancho and Vaughan (2020). This package is geared towards time series analysis, and it can be performed on remainders with removed seasonal and trend components similar to the method mentioned. Next, we illustrate how to use the STL() function in the "anomalize" package to detect anomalies in time series.

```
#library(devtools)
#devtools::install_github('business-science/anomalize')
```

```
library(anomalize)
library(tidyverse)
```

There are three main functions in a typical workflow for anomaly detection:

- `time_decompose()`: separates the time series into seasonal, trend, and remainder components.

- `anomalize()`: applies anomaly detection methods to the remainder component.

- `time_recompose()`: calculates limits that separate the "normal" data from the anomalies.

The following code demonstrates a typical example of how to use these functions to detect anomalies for time series.

```
NewJersey.ts %>%
  # Separate it into seasonal, trend, and remainder components
  time_decompose(Y.Death, method = "stl") %>%
  # Apply anomaly detection to the remainder
  anomalize(remainder, alpha = 0.05, max_anoms = 0.05) %>%
  time_recompose() %>%
  # Anomaly visualization
  plot_anomalies(time_recomposed = TRUE, ncol = 3,
                 alpha_dots = 0.25) +
  labs(title = "Tidyverse Anomalies",
       subtitle = "alpha = 0.05, max = 5%")
```

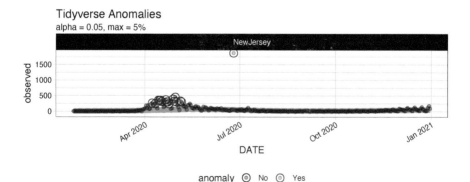

FIGURE 9.37: Tidyverse anomalies from the daily new deaths in New Jersey.

In the above, the `alpha` and `max_anoms` are the two parameters that control the `anomalize()` function. By default, the `alpha` is set to 0.05, and when we decrease the value of `alpha`, the band will increase, which will reduce the chance for an outlier to occur. In practice, `alpha` is recommended to be chosen from 0.02 to 0.10. The parameter `max_anoms` controls the maximum percentage of data that can be an anomaly. This is useful when `alpha` is too difficult to tune, and we can set the value of `max_anoms` to focus on the most severe anomalies.

9.10.5   A Discussion on Outlier and Anomalies Repair

In general, simply replacing or removing the outliers and the anomalies without thinking about the reason for them to occur is a dangerous practice. They may be informative about the data-producing process and should be considered when forecasting. For example, the anomaly in Figure 9.37 is due to New Jersey reporting 1,854 probable deaths that may date back to earlier in the outbreak. We suggest that data analysts review the detected outliers and/or anomalies individually and investigate the reasons for the outliers and/or anomalies to determine whether a correction is appropriate.

Correcting the history for severe outliers or anomalies will sometimes improve the forecast; however, it might lead to poor forecasts if the correction is unnecessary. In practice, we find that if the outlier is not genuinely severe, corrections might make the history smoother than it actually was, which will change the forecasts and narrow the prediction intervals. Wang et al. (2021a) recommended using a high threshold for anomaly detection. In certain cases, infectious disease data analysts need to work with epidemiologists to manually establish static thresholds for each monitored metric.

9.11   Further Reading

For the packages introduced in this chapter, there are valuable web resources available:

- `https://fable.tidyverts.org/;`
- `https://tidyverts.github.io/tidy-forecasting-principles/;`
- `https://cran.r-project.org/web/packages/anomalize/index.html;`
- `https://www.r-pkg.org/pkg/anomalize.`

For good reference books on the time series forecasting topics covered, see:

- Hyndman, R.J. and Athanasopoulos, G. (2021). *Forecasting: principles and practice*. 3rd edition, OTexts: Melbourne, Australia. Available at `https://otexts.com/fpp3/`.

- Brockwell, P. J. and Davis, R. A. (2002). *Introduction to time series and forecasting*. New York, NY: Springer New York.

9.12 Exercises

1. From `state.long` in the R package IDDA, choose a state and convert the data to time series.

   (a) Construct time series plots for the daily new death, daily new infection, cumulative deaths and cumulative infections.
   (b) For the daily new death series and daily new infection series, use the following graphics functions: `autoplot()`, `gg_season()`, `gg_subseries()`, `gg_lag()`, `ACF()` and explore features from the following time series:
      - Can you spot any seasonality and trend?
      - What do you learn about the series?
      - What can you say about the seasonal patterns?
      - Can you identify any unusual pattern?
   (c) Conduct an STL decomposition for the daily new death series.
   (d) For the daily new death series, create a training dataset consisting of observations from January 23 to December 17, 2020. Calculate 14-day-ahead forecasts using `SNAIVE()` and `ETS` applied to your training data.
   (e) Compare the accuracy of your forecasts against the actual values.
   (f) Check the residuals. Do the residuals appear to be uncorrelated and normally distributed?

2. Let $\{Z_t\}$ be a sequence of independent normal random variables, each with mean 0 and variance σ^2, and let a, b, and c be constants. Which of the following processes, if any, is stationary? For each stationary process, please specify the mean and autocovariance function.

   (a) $Y_t = a + bZ_t + cZ_{t-2}$
   (b) $Y_t = Z_1 \cos(ct) + Z_2 \sin(ct)$
   (c) $Y_t = Z_t \cos(ct) + Z_{t-1} \sin(ct)$
   (d) $Y_t = a + bZ_0$

(e) $Y_t = Z_0 \cos(ct)$
(f) $Y_t = Z_t Z_{t-1}$

3. Use the `tsibble()` function to generate a time series of length 200 of the following simple ARIMA models. Produce a time plot for each simulated series, and draw the sample ACF and PACF for the simulated time series.

   (a) AR(1) model with $\phi_1 = 0.9$ and $\sigma^2 = 1$.
   (b) MA(1) model with $\theta_1 = 0.8$ and and $\sigma^2 = 1$.
   (c) MA(2) model with $\theta_1 = 0.3$, $\theta_2 = -0.4$ and $\sigma^2 = 1$.

4. From `state.long` in the R package IDDA, choose a state. Hold out the last seven days as test data and use all the data except the last seven days as your training data. Use the automatic search in `ARIMA()` to find an ARIMA model for the daily new deaths based on your training data. What model was selected? Write the model in terms of the backshift operator.

5. For the state you choose in Problem 4, consider a logarithm transformation, then use the automatic search in `ARIMA()` to find an ARIMA model for the daily new deaths based on your training data. Make a 7-day-ahead forecast using the selected ARIMA model, and provide 95% prediction intervals.

6. Compare the forecasting performance of the selected ARIMA model in Problem 4 and Problem 5 without and with logarithm transformation.

7. For the state you choose in Problem 4, and the ARIMA model for the daily new deaths with log transformation, aggregate them to make a 7-day-ahead forecast for the cumulative number of deaths and provide the corresponding 95% prediction intervals.

10

Regression Methods

In Chapter 9, we introduced the time series analysis tools, which work on an ordered series of data to forecast future patterns. These tools treat the value of a target variable at each time point as a random variable, then the correlation among these random variables is presented using their covariance function. Another popular approach is regression methods, which can also be applied to time-series problems, such as auto-regression models. Furthermore, regression techniques can be applied to non-ordered data, too. In general, in regression, we usually assume the output variable/response variable depends on the values taken by other explanatory variables/covariates. Then the new values of these covariates are used to make predictions of the response variable. This chapter will focus on the latter approach.

Starting with investigating the relationship between the explanatory and outcome variables, this chapter first reviews parametric methods that can be applied to quantify the effect of a set of explanatory variables of a particular outcome. Next, we introduce several nonparametric regression methods in which there is no parametric form assumed for the relationship between the outcome and the explanatory variables. Instead, the nonparametric methods construct the smooth function according to the information derived from the data. Later, several dynamic updating methods are proposed to improve point forecast accuracy. Last but not least, several generalized linear regression models are introduced to take care of the discrete outcome variables of interest in epidemiology, such as a count of disease events for area units or a binary variable indicating whether the disease is present at a given location.

10.1 Parametric Regression Methods

Regression analysis traces the average value of a response variable (also known as a dependent variable) as a function of one or several explanatory variables (also known as covariates, predictors, or independent variables). A parametric regression model will first select a form for the function and then estimate the parameters of the function from the data. Parametric models have several advantages: (i) they are easy to understand and interpret; (ii) they are usually computationally fast, meaning they learn fast from data; and (iii) they can handle data with relatively small sample sizes and still give decent results.

10.1.1 Linear Regression and Nonlinear Regression

Linear regression attempts to model the relationship between the response variable (Y) and explanatory variables (X's) by fitting a linear equation to observed data. Given n observations $\{(Y_i, X_{i1}, \ldots, X_{id}), i = 1, \ldots, n\}$, multiple linear regression allows the mean value of a continuous response variable Y_i to be represented as a function of d explanatory variables X_{ij}, $j = 1, \ldots, d$:

$$Y_i = \beta_0 + \beta_1 X_{i1} + \beta_2 X_{i2} + \cdots + \beta_d X_{id} + \epsilon_i,$$

where $\beta_0, \beta_1, \ldots, \beta_d$ are unknown parameters or coefficients, and ε_i is the error term.

Usually, there are four key assumptions behind this type of regression analysis: (i) the relationship between the response and regressors is linear, at least approximately; (ii) the errors all have the same variance, i.e., $\text{Var}(\epsilon_i) = \sigma^2$ for any i; (iii) the errors are independent of each other, i.e., the value of the errors at any point is not affected by the value of the errors at any other point; (iv) the residuals ϵ_i are normally distributed with a mean of zero, i.e., $\text{E}(\epsilon_i) = 0$ for any i.

Nonlinear regression is a method to model a nonlinear relationship between the response variable and one or a set of explanatory variables, which takes the form:

$$Y_i = m(X_i, \beta) + \epsilon_i,$$

where X_i is a vector of d predictors, β is a vector of k parameters, m is a known regression function, ϵ is an error term. For example, the Michaelis-Menten model is a nonlinear model with the following nonlinear regression function

$$m(x, \beta) = \frac{\beta_1 x}{\beta_2 + x}.$$

Other examples of nonlinear functions include exponential functions, logarithmic functions, power functions, and trigonometric functions.

Linear and nonlinear models are two classes of parametric regression models since the function that describes the relationship between the outcome and explanatory variables is assumed to be known. Linear regression estimates the coefficients from the data using the least-squares method that minimizes the norm of a residual vector. The estimation for a nonlinear model is usually achieved by some search methods from the optimization.

Example 10.1. (Modeling and Forecasting COVID-19 County-level Death Count Using Parametric Regression Methods). To predict the number of deaths for COVID-19, Altieri et al. (2021) considered the following regression models for modeling the death count at the county level in the US, in which the future death count is assumed to follow a linear or exponential relationship with time or the current death count. Let $D_{c,t}$ be the death count on day t for county c, $c = 1, ..., C$. Below, we introduce the five regression models proposed by Altieri et al. (2021).

A separate-county exponential predictor: uses an exponential function of time to predict the death counts for each county, that is,

$$E(D_{c,t+1}|t) = \exp\left\{\beta_0^c + \beta_1^c(t+1)\right\}, \ c = 1, ..., C. \tag{10.1}$$

A separate-county linear predictor: similar to the separate county exponential predictor, but instead of using the exponential format, a linear format is considered, that is,

$$E(D_{c,t+1}|t) = \beta_0^c + \beta_1^c(t+1), \ c = 1, ..., C. \tag{10.2}$$

A shared-county exponential predictor: uses an exponential function of the death counts in the past, $D_{c,t}$, to predict death counts, that is,

$$E(D_{c,t+1}|t) = \exp\left\{\beta_0 + \beta_1 \log\left(D_{c,t} + 1\right)\right\}, \ c = 1, ..., C. \tag{10.3}$$

An expanded shared-county exponential predictor: similar to the shared-county exponential predictor but also includes counts of COVID-19 cases, $I_{c,t}$, and neighboring county cases, $I_{c,t}^{\text{neighbor}}$, and deaths, $D_{c,t}^{\text{neighbor}}$, as predictive features. The model strives to incorporate spatial information and effects of previous cases, up to the end of day $t - k + 1$, associated with the future death:

$$
\begin{aligned}
E(D_{c,t+1}|t) = {} & \exp\left\{\beta_0 + \beta_1 \log\left(D_{c,t} + 1\right) + \beta_2 \log\left(I_{c,t-k+1} + 1\right)\right. \\
& \left. + \beta_3 \log\left(D_{c,t-k+1}^{\text{neighbor}}\right) + \beta_4 \log\left(I_{c,t-k+1}^{\text{neighbor}}\right)\right\},
\end{aligned} \tag{10.4}
$$

for $c = 1, ..., C$.

A demographics shared-county exponential predictor: a predictor also similar to the shared-county exponential predictor but with additional covariates to address local features for each county such as county density and size, demographic information, and health care resources, d_j^c, $j = 1, ..., m$, that is,

$$E(D_{c,t+1}|t) = \exp\left\{\beta_0 + \beta_1 \log\left(D_{c,t} + 1\right) + \sum_{j=1}^{m} \gamma_j d_j^c\right\}, \ c = 1, ..., C. \tag{10.5}$$

Next, we illustrate how to use models in (10.1) and (10.2) to make predictions based on the data `CA.county.ts`, which records the daily COVID-19 death count for Los Angeles County, CA. Suppose we are interested in making predictions for the daily new death count (`Y.Death`). We use a training set from November 25 to December 4, 2020, to predict the period from December 5 to December 11, 2020. The linear regression model and exponential model will be fitted on the training set, and the fitted model is used to make predictions for the future count. Figure 10.1 shows the forecasts of the daily death count.

```
# Install and load the packages
library(dplyr); library(fable); library(tsibble)
library(feasts); library(ggplot2); library(tidyr)
# devtools::install_github('FIRST-Data-Lab/IDDA')
library(IDDA)
data(CA.county.ts)

# Daily new deaths in Los Angeles County, CA
LA.ts <- CA.county.ts %>%
  dplyr::filter(County == "LosAngeles") %>%
  dplyr::select(Death, Y.Death)



# Set training data from November 25 to December 4
train <- LA.ts %>%
  filter_index("2020-11-25" ~ "2020-12-04")
n <- nrow(train)

# Obtain both linear and exponential trend
fit_trends <- train %>%
  model(
    linear = TSLM(Y.Death ~ trend()),
    exponential = TSLM(log(Y.Death + 1) ~ trend()),
  )
fc_trends <- fit_trends %>% forecast(h = 7)

# Make predictions for the next week
LA.ts %>%
  dplyr::filter(DATE < train$DATE[n] + 7) %>%
  autoplot(Y.Death) +
  geom_line(data = fitted(fit_trends),
            aes(y = .fitted, color = .model)) +
  autolayer(fc_trends, alpha = 0.5, level = 95) +
```

```
labs(y = "Number of deaths",
     title = "Daily new deaths in Los Angeles County, CA") +
theme_bw() +
theme(legend.position = "bottom")
```

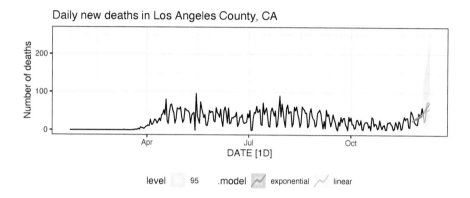

FIGURE 10.1: Reported death count with linear and exponential prediction.

10.1.2 Model Adequacy Checking

After model fitting, we should consider the validity of the aforementioned assumptions in Section 10.1. Violations of the assumptions may yield an unstable model in the sense that a different sample could lead to a different model with opposite conclusions. Statistics, such as the coefficient of determination, can be used to check the linearity assumption between explanatory and response variables. Graphical analysis of residuals (original or scaled) is a very effective way to investigate the adequacy of the fit. These include the normal probability plot of residuals, the plot of residuals against the fitted values, the plot of residuals against each regressor variable, and the plot of residuals in time series if time series data were collected.

10.1.2.1 Goodness of Fit

How well does the model fit the data? One measure is R^2, the so-called **coefficient of determination** or **percentage of variance explained**,

$$R^2 = 1 - \frac{\Sigma(Y_i - \hat{Y}_i)^2}{\Sigma(Y_i - \bar{Y}^2)} = 1 - \frac{\text{Residual Sum of Squares}}{\text{Total Sum of Squares (corrected for mean)}}.$$

The range is $0 \leq R^2 \leq 1$, and values closer to 1 indicate better fits. For simple linear regression, $R^2 = r^2$, where r is the correlation between the explanatory variable X and the response variable Y. An equivalent definition is

$$R^2 = \frac{\sum (\hat{Y}_i - \bar{Y})^2}{\sum (Y_i - \bar{Y})^2} = \frac{\text{Regression Sum of Squares}}{\text{Total Sum of Squares (corrected for mean)}}.$$

This coefficient allows us to select a set of variables that well explains the variability in Y. In practice, it is important to plot the residuals after choosing the regression variables and fitting a model to check whether the model's assumptions have been satisfied.

10.1.2.2 ACF Plot of Residuals

For time series data, the value of a variable observed in the current time period might be similar to its value in the past, such as the values observed in the previous period or even the period before that. Therefore, autocorrelation is commonly found among the residuals when fitting a regression model to time series. In this case, the assumption of no autocorrelation in the errors cannot be satisfied, and the forecasts may be inefficient, which usually means some information that should be accounted for in the model is left. However, the forecasts from a model with autocorrelated errors are still unbiased and not "wrong," but they will usually have wider prediction intervals than they need to. Therefore we should always look at an ACF plot of the residuals to check the independence assumption of the errors.

10.1.2.3 Histogram and Normal Quantile Plot of Residuals

Checking whether the residuals are normally distributed is always a good idea. Although it might not be essential for forecasting, it does improve the efficiency of the prediction intervals.

10.1.2.4 Residual Plot against Predictors

Typically, we would expect the residuals to be randomly scattered and to not show any systematic patterns. To check this, scatterplots of the residuals against each explanatory variable are a helpful and straightforward way. If a pattern is shown in these scatterplots, then the relationship may be nonlinear (or the variance of residual is not constant across time), and thus we need to modify the model accordingly.

In addition, it is also necessary to plot the residuals against any other explanatory variables that are not in the model. If a pattern is shown in any of these plots, then we may need to consider adding the corresponding explanatory variable to the model, possibly in a nonlinear form.

10.2 Nonparametric Regression Methods

In practice, parametric models are usually suited for simple problems due to the restriction of the function form. **Nonparametric regression** offers a flexible alternative to parametric methods for regression by "letting the data speak for themselves." Unlike parametric methods, which assume that the regression relationship has a known form that depends on a finite number of unknown parameters, nonparametric regression models attempt to learn the form of the regression relationship from a sample of data.

A nonparametric regression model assumes

$$Y_i = m(X_i) + \varepsilon_i, \quad i = 1, \dots, n, \tag{10.6}$$

where $m(\cdot)$ is an unknown function, and ε_i are assumed to be i.i.d. with mean 0 and variance σ^2.

For time series data, we have introduced how to find the trend using parametric methods. We can also estimate the trend nonparametrically by assuming that

$$Y_t = m(t) + \varepsilon_t, \quad t = 1, \dots, n,$$

where $m(\cdot)$ is a nonparametric regression function. Linear regression is a special case:

$$m(t) = \beta_0 + \beta_1 t.$$

To estimate the function $m(\cdot)$ in (10.6) nonparametrically, we will introduce the spline smoothing method. Splines are a popular family of smoothers, which are practically as easy to implement and fast as a simple linear regression with a slowly increasing number of parameters (Ruppert et al. (2003)). Below we will introduce several spline smoothing techniques and illustrate how they can be used for trend estimation in time series analysis.

10.2.1 Piecewise Constant Splines

We break the time domain into bins, and fit a different constant in each bin. We create cut-off points $k_1, k_2, \dots, k_N$ in the time domain, and then construct $N + 1$ piecewise constant basis functions:

$$
\begin{aligned}
B_0(t) &= I(t < k_1), \\
B_1(t) &= I(k_1 \le t < k_2), \\
B_2(t) &= I(k_2 \le t < k_3), \\
\cdots &= \cdots, \\
B_{N-1}(t) &= I(k_{N-1} \le t < k_N), \\
B_N(t) &= I(k_N \le t),
\end{aligned}
$$

where $I(\cdot)$ is an indicator function that returns a 1 if the condition is true, and returns a 0 otherwise.

Example 10.2. (Trend Estimation of Florida COVID-19 Death Count Time Series Using Piecewise Constant Splines). We are interested in estimating the trend for the daily new death count time series for Florida. The time series data can be obtained from the state.ts in the IDDA package.

The variable Y.Death contains the time series of the daily new death count for Florida from January 22 to December 31, 2020. The following code shows how to obtain a piecewise constant spline fit for the trend of Y.Death.

Figure 10.2 shows a fit for the daily new death count for the state of Florida using piecewise constant splines.

```r
n <- nrow(Florida.ts)
t <- 1:n
y <- Florida.ts$Y.Death

# Knots
N <- 21
knots <- 1 + (n-1)/(N+1) * (0:N)

# Piecewise constant spline basis
t.rep <- matrix(rep(t, N), n, N)
knot.L <- matrix(rep(knots[-(N + 1)], each = n), n, N)
knot.R <- matrix(rep(knots[-1], each = n), n, N)
B <- 1*((knot.L <= t.rep) & (t.rep < knot.R))
X <- cbind(B, knots[N] < t & t <= n)

# Piecewise constant spline fit
M <- t(X) %*% X
beta <- solve(M) %*% t(X) %*% y
yhat <- X %*% beta
Florida.ts$pcs_preds <- yhat



# Plot of reported vs piecewise constant spline fit
Florida.ts %>%
    ggplot(aes(x = DATE)) +
    geom_line(aes(y = Y.Death, color = "Reported")) +
    geom_line(aes(y = pcs_preds, color = "Fitted")) +
    scale_color_manual(
```

```
    values = c(Reported = "black", Fitted = "red")
  ) +
  labs(y = "Number of deaths",
       title = "Reported vs piecewise constant spline fit") +
  guides(color = guide_legend(title = "Series")) +
 theme(legend.position = "bottom")
```

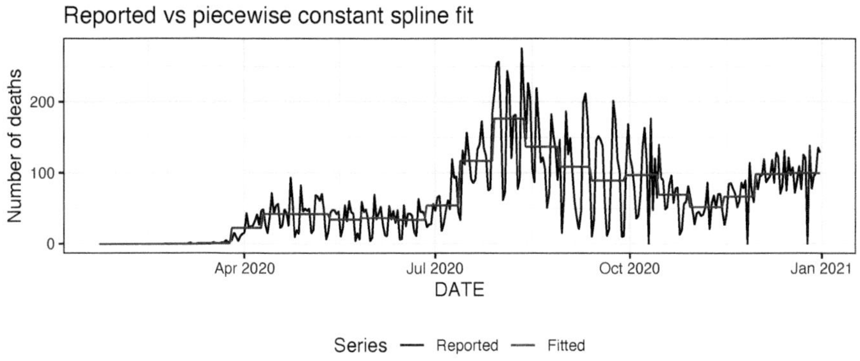

FIGURE 10.2: Piecewise constant spline smoothing for the daily new death count for Florida.

10.2.2 Truncated Power Splines

For univariate data, polynomial splines of degree p can be represented by an appropriate sequence of $N + p + 1$ spline basis functions, determined in turn by N interior knots. These produce functions that are piecewise polynomials of degree p between the knots and joined up with the continuity of degree $p - 1$ at the knots. As an example, we consider linear splines or piecewise linear functions.

Let $k_1 < k_2 < ... < k_N$ be **knots** in the time domain, for example, $[1, n]$. Let $\phi_0(t) = 1$, $\phi_1(t) = t$, $\phi_j(t) = (t - k_{j-1})_+$ for $j = 2, ..., N + 1$ be the basis functions, where x_+ denotes the positive part of x. Then, the regression is piecewise linear with bends at the knots.

In general, let

$$(t - k_j)_+^p = \begin{cases} (t - k_j)^p & \text{for } t \ge k_j \\ 0; & \text{for } t < k_j. \end{cases}$$

Then, the $N + p + 1$ truncated power basis functions are:

$$
\begin{aligned}
\phi_1(t) &= 1, \\
\phi_2(t) &= t, \\
&\vdots \\
\phi_p(t) &= t^p, \\
\phi_{p+1}(t) &= (t - k_1)_+^p, \\
&\vdots \\
\phi_{p+N}(t) &= (t - k_N)_+^p.
\end{aligned}
$$

We define the spline estimator for the regression function $m(x)$ as

$$
\hat{m}(t) = \sum_{k=0}^{p} \hat{\beta}_k t^k + \sum_{j=1}^{N} \hat{\gamma}_j \left(t - \kappa_j \right)_+^p,
$$

where $\{\hat{\beta}_0, \dots, \hat{\beta}_p, \hat{\gamma}_1, \dots, \hat{\gamma}_N\}$ are the least squares estimators of $\{\beta_0, \dots, \beta_p, \gamma_1, \dots, \gamma_N\}$ based on the data. The shape of the basis functions is determined by the position of the knots $k_1 < \dots < k_N$, which can, for example, be uniformly spread over the time domain.

Example 10.3. (Trend Estimation of Florida COVID-19 Death Count Time Series Using Truncated Power Splines). Similar to Example 10.2, we are interested in estimating the trend for the daily new death count time series for Florida, but we will use the truncated power splines this time. The result of prediction can be found in 10.3.

```
y <- Florida.ts$Y.Death
n <- nrow(Florida.ts)
t <- 1:n


# Knots
N <- 10
knots <- 1 + (n-1)/(N+1) * (0:N)


# Truncated power spline basis functions
X <- matrix(1, n, N + 2)
X[, 2] <- t
t.rep <- matrix(rep(t, N), n, N)
tmp <- t.rep - matrix(rep(knots[2:(N + 1)], each = n), n, N)
X[, 3:(N+2)] <- tmp * (tmp > 0)


# Truncated power spline fit
```

```
M <- t(X) %*% X
beta <- solve(M) %*% t(X) %*% y
yhat <- X %*% beta
Florida.ts$tps_preds <- yhat

# Plot of reported vs truncated power spline fit
Florida.ts %>%
    ggplot(aes(x = DATE)) +
    geom_line(aes(y = Y.Death, color = "Reported")) +
    geom_line(aes(y = tps_preds, color = "Fitted")) +
    scale_color_manual(
      values = c(Reported = "black", Fitted = "red")
    ) +
    labs(y = "Number of deaths",
          title = "Reported vs truncated power spline fit") +
    guides(color = guide_legend(title = "Series")) +
    theme(legend.position = "bottom")
```

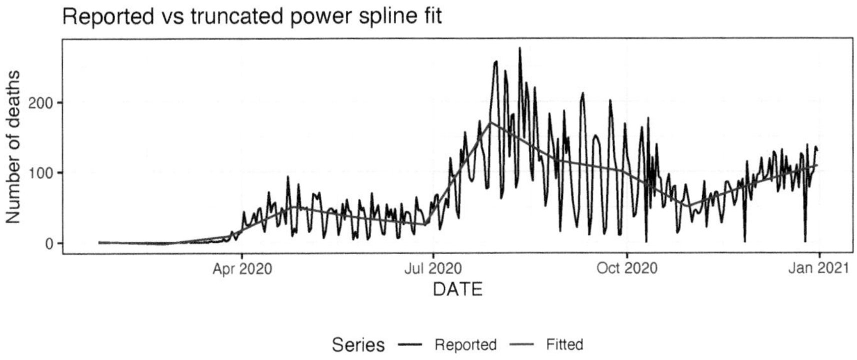

FIGURE 10.3: Truncated power spline smoothing for the daily new death count for Florida.

10.2.3 B-splines and Natural Splines

B-splines and natural splines similarly define a basis over the time domain. They are made up of piecewise polynomials of a given degree and have defined derivatives similar to the piecewise-defined functions.

To introduce the space of splines, we pre-select an integer N, and divide the time domain $[a, b]$ into $(N + 1)$ subintervals $J_j = [k_j, k_{j+1})$, $j = 0, ..., N - 1$,

$J_N = [k_N, b]$, where $\{k_j\}_{j=1}^N$ is a sequence of equally spaced points, called interior knots, given as

$$k_{1-p} = \ldots = k_{-1} = k_0 = a < k_1 < \ldots < k_N < b = k_{N+1} = \ldots = k_{N+p},$$

in which $k_j = j(b-a)/(N+1)$, $j = 0, 1, \ldots, N+1$. The j-th B-spline of order r denoted by $b_{j,r}$ is recursively defined by De Boor (1978) as follows:

- For $r = 1$, basis $b_{1,1}, \ldots, b_{N,1}$:

$$b_{j,1}(t) = I \left\{ t \in [k_j, k_{j+1}) \right\}.$$

- Given $b_{j,r-1}, j = -(r-1), \ldots, N$, construct $b_{j,r}$

$$b_{j,r}(t) = \frac{t - k_j}{k_{j+r-1} - k_j} b_{j,r-1}(t) + \frac{k_{j+r} - t}{k_{j+r} - k_{j+1}} b_{j+1,r-1}(t)$$

- E.g., $r = 2$, basis $b_{-1,2}, \ldots, b_{N,2}$:

$$b_{j,2}(t) = \frac{t - k_j}{k_{j+1} - k_j} b_{j,1}(t) + \frac{k_{j+2} - t}{k_{j+2} - k_{j+1}} b_{j+1,1}(t).$$

It is well known that the behavior of polynomial fit tends to be erratic near the boundaries, and extrapolation can be dangerous. These problems are exacerbated with splines. Beyond the boundary knots, the polynomials fit usually behaves even more wildly than the corresponding global polynomials. This can be conveniently summarized via the pointwise variance of spline functions fit by least squares.

A **natural cubic spline** adds additional constraints, namely that the function is linear beyond the boundary knots. This frees up four degrees of freedom (two constraints each in both boundary regions), which can be spent more profitably by sprinkling more knots in the interior region.

Example 10.4. (Trend Estimation of Florida COVID-19 Death Count Time Series Using Natural Splines). This example shows how to estimate the trend for the daily new death count time series for Florida using the natural cubic splines.

```
# Natural spline fit
library(splines)
n <- nrow(Florida.ts)
t <- 1:n
ns_fit <- lm(Y.Death ~ ns(t, df = 6), data = Florida.ts)
summary(ns_fit)
```

```
##
## Call:
## lm(formula = Y.Death ~ ns(t, df = 6), data = Florida.ts)
##
## Residuals:
##      Min      1Q  Median      3Q     Max
## -117.47  -16.66   -1.87   19.97  149.75
##
## Coefficients:
##                   Estimate Std. Error t value Pr(>|t|)
## (Intercept)         -13.3      10.4   -1.28    0.201
## ns(t, df = 6)1       29.5      13.1    2.25    0.025
## ns(t, df = 6)2      107.8      16.8    6.43  4.4e-10
## ns(t, df = 6)3      179.6      14.9   12.06   < 2e-16
## ns(t, df = 6)4       14.9      13.0    1.15    0.251
## ns(t, df = 6)5      134.7      26.4    5.10  5.8e-07
## ns(t, df = 6)6      116.7      11.9    9.82   < 2e-16
##
## (Intercept)
## ns(t, df = 6)1 *
## ns(t, df = 6)2 ***
## ns(t, df = 6)3 ***
## ns(t, df = 6)4
## ns(t, df = 6)5 ***
## ns(t, df = 6)6 ***
## ---
## Signif. codes:
## 0 '***' 0.001 '**' 0.01 '*' 0.05 '.' 0.1 ' ' 1
##
## Residual standard error: 39.6 on 337 degrees of freedom
## Multiple R-squared:  0.548,  Adjusted R-squared:  0.539
## F-statistic:   68 on 6 and 337 DF,  p-value: <2e-16


Florida.ts$ns_preds <- predict(ns_fit)

# Natural spline prediction and prediction intervals
h <- 14
t.new <- t[n] + (1:h)
ns_PI <- predict(ns_fit, newdata = data.frame(t = t.new),
                 interval = "prediction", level = 0.95)

ns_PI <- as.data.frame(ns_PI) %>%
  mutate(DATE = (Florida.ts$DATE)[n] + 1:h)
```

```
# Plot of reported vs natural spline fit
ns <- Florida.ts %>%
    ggplot(aes(x = DATE)) +
    geom_line(aes(y = Y.Death, color = "Reported")) +
    geom_line(aes(y = ns_preds, color = "Fitted")) +
    scale_color_manual(
      values = c(Reported = "black", Fitted = "red")
    ) +
    labs(y = "Number of deaths",
         title = "Reported vs natural spline regression fit") +
    guides(color = guide_legend(title = "Series")) +
    theme(legend.position = "bottom")

# Plot of natural spline fit and its prediction intervals
ns_int <- ns +
    geom_ribbon(
      mapping = aes(y = fit,
                    ymin = lwr,
                    ymax = upr,
                    fill = '95% Prediction Intervals'),
            data = ns_PI, alpha = 0.2) +
    geom_line(mapping = aes(y = fit, color = "Fitted"),
            data = ns_PI,
            key_glyph = "timeseries") +
    labs(title = "Natural spline regression fit and prediction
      ↵  intervals")  +
    guides(color = guide_legend(title = "Series"),
           fill =  guide_legend(title = "")) +
    theme(legend.position = "bottom")
```

10.2.4   Smoothing Splines

A different idea is to estimate the regression function $m(\cdot)$ by

$$\min_{m} \frac{1}{n} \sum_{t=1}^{n} \{Y_t - m(X_t)\}^2 + \lambda \int \{m''(t)\}^2 dt,$$

where $\lambda > 0$ is a smoothing parameter, and it controls large values of the second derivative of m. For $\lambda = 0$, no penalty is imposed, and any interpolating function will do, while for $\lambda = \infty$, only functions linear in t are permitted.

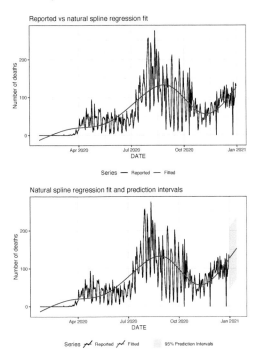

FIGURE 10.4: Plot for the daily new death count for Florida. Top: reported values vs fitted values using natural splines. Bottom: natural spline fit and its prediction intervals.

Example 10.5. (Trend Estimation of Florida COVID-19 Death Count Time Series Using Smoothing Splines). This example illustrates how to estimate the trend for Florida's daily new death count time series using smoothing splines.

The gam() function in the R package "mgcv" can implement the smoothing in a highly automatic manner. With the "mgcv" package, spline smooths can be included in model formulae of the gam() using the s() specification.

```
# Install and load the package
if(!require('mgcv')) install.packages('mgcv')
library(mgcv)
```

```
# Smoothing spline fit
ss_fit <- gam(Y.Death ~ s(t, bs = "cr"), data = Florida.ts)
summary(ss_fit)
```

```
##
## Family: gaussian
## Link function: identity
##
## Formula:
## Y.Death ~ s(t, bs = "cr")
##
## Parametric coefficients:
##              Estimate Std. Error t value Pr(>|t|)
## (Intercept)     63.00       1.94    32.5   <2e-16 ***
## ---
## Signif. codes:
## 0 '***' 0.001 '**' 0.01 '*' 0.05 '.' 0.1 ' ' 1
##
## Approximate significance of smooth terms:
##        edf Ref.df     F p-value
## s(t) 8.79   8.99  62.8  <2e-16 ***
## ---
## Signif. codes:
## 0 '***' 0.001 '**' 0.01 '*' 0.05 '.' 0.1 ' ' 1
##
## R-sq.(adj) =   0.62   Deviance explained =   63%
## GCV = 1329.3  Scale est. = 1291.5    n = 344
```

```
Florida.ts$ss_preds <- predict(ss_fit)

# Plot of reported vs smoothing spline fit
Florida.ts %>%
    ggplot(aes(x = DATE)) +
    geom_line(aes(y = Y.Death, color = "Reported")) +
    geom_line(aes(y = ss_preds, color = "Fitted")) +
    scale_color_manual(
      values = c(Reported = "black", Fitted = "red")) +
    labs(y = "Number of deaths",
         title = "Reported vs smoothing spline fit") +
```

```
guides(color = guide_legend(title = "Series"))   +
theme(legend.position = "bottom")
```

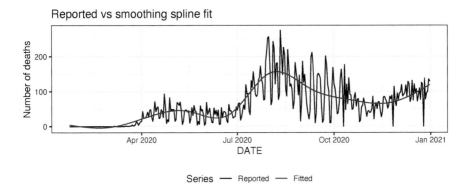

FIGURE 10.5: Plot for the daily new death count for Florida: reported values (black) and fitted values (red) using smoothing splines.

10.3   An Application to CDC FluView Portal Data

The CDC FluView Portal provides the national, regional, and state-level outpatient illness and viral surveillance data in-season and past seasons from both ILINet (Influenza-like Illness Surveillance Network) and WHO/NREVSS (National Respiratory and Enteric Virus Surveillance System).

"FluView" provides a weekly influenza surveillance report, and "FluView" Interactive is an online application that is updated each week. It can be used for a more in-depth exploration of influenza surveillance data.

R Package "cdcfluview" retrieves Flu Season Data from the CDC "FluView" Portal. We can use the function ilinet() to retrieve state, regional, or national influenza statistics from the CDC.

```
library(cdcfluview); library(dplyr)
library(tsibble); library(lubridate)

# Prepare the data
```

```
usflu.raw <- ilinet("national", years = 2010:2018)
names(usflu.raw)
```

```
##  [1] "region_type"        "region"
##  [3] "year"               "week"
##  [5] "weighted_ili"       "unweighted_ili"
##  [7] "age_0_4"            "age_25_49"
##  [9] "age_25_64"          "age_5_24"
## [11] "age_50_64"          "age_65"
## [13] "ilitotal"           "num_of_providers"
## [15] "total_patients"     "week_start"
```

```
usflu <- usflu.raw %>%
  mutate(
    date = as.Date(paste0(year, sprintf("%02d", week), "00"),
                   format="%Y%W%w"),
    dec_date = decimal_date(week_start),
    week = yearweek(week_start),
    time_in_year = dec_date%%1) %>%
  dplyr::filter(!is.na(dec_date))

usflu.ts <- as_tsibble(usflu, index = week)
```

In the following, we will consider the variable weighted_ili as the response, the percentage of outpatient doctor visits for influenza-like illness, weighted by state population. The top panel of Figure 10.6 displays the weekly time series of weighted_ili from 2010 to 2019.

```
usflu.ts %>% autoplot(weighted_ili) +
    labs(x = "week",
         title = "National influenza-like illness") +
    guides(color = guide_legend(title = "Series")) +
    theme_bw()
```

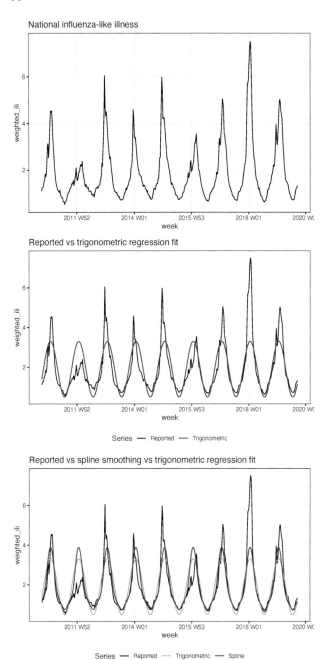

FIGURE 10.6: Top: national influenza-like illness weekly time series. Center: reported vs trigonometric regression fit. Bottom: spline smoothing vs trigonometric regression fit.

10.3.1   Trigonometric Regression

For the ith year and jth reported time point t_{ij}, $i = 1, 2, ... , n$, let Y_{ij} be the measurement of variable `weighted_ili`. Suppose that

$$Y_{ij} = \beta_0 + \beta_1 \sin(2\pi t_{ij}) + \beta_2 \cos(2\pi t_{ij}) + \varepsilon_{ij},$$

where $t_{ij} \in [0, 1]$, β_0 and β_1 are unknown coefficients. We will consider the linear regression method to estimate β_0 and β_1.

```
trig_fit <- lm(weighted_ili ~ sin(time_in_year*2*pi)
               + cos(time_in_year*2*pi),
               data = usflu)
usflu$trig_preds <- predict(trig_fit)
```

The middle panel of Figure 10.6 displays the reported and trigonometric regression fit of `weighted_ili` from 2010 to 2019.

```
ggplot(usflu, aes(x = week)) +
    geom_line(aes(y = weighted_ili, color = "Reported")) +
    geom_line(aes(y = trig_preds, color = "Trigonometric")) +
    scale_color_manual(
      values = c(Reported = "black", Trigonometric = "red")
    ) +
    labs(x = "week",
         title = "Reported vs trigonometric regression fit") +
    guides(color = guide_legend(title = "Series"))
```

10.3.2   Smoothing Splines

For the ith year and jth reported time point t_{ij}, we assume that the response variable (`weighted_ili`) Y_{ij} satisfies the following nonparametric regression model:

$$Y_{ij} = m(t_{ij}) + \varepsilon_{ij},$$

where $t_{ij} \in [0, 1]$, $i = 1, 2, ... , n$. In the following, we use the smoothing splines to estimate the $m()$ function.

```
library(mgcv)
# Fit a GAM model using smoothing splines
ss_fit <- gam(weighted_ili ~ s(time_in_year, bs = "cc"),
```

```
                data = usflu)
usflu$ss_preds <- predict(ss_fit)
```

The bottom panel of Figure 10.6 displays the reported weighted_ili and the fitted weighted_ili from 2010 to 2019 based on the trigonometric regression method and spline smoothing method.

```
ggplot(usflu, aes(x = week)) +
    geom_line(aes(y = weighted_ili, color = "Reported")) +
    geom_line(aes(y = trig_preds, color = "Trigonometric")) +
    geom_line(aes(y = ss_preds, color = "Spline")) +
    scale_color_manual(
       values = c(Reported = "black", Trigonometric = "cyan3",
                  Spline = "red")) +
    labs(x = "week",
         title = "Spline smoothing vs trigonometric regression fit") +
    guides(color = guide_legend(title = "Series"))
```

10.4 Poisson Regression

In infectious disease analysis, the outcome of interest is often a count of the number of events occurring in a population of a given size or a count of the number of events concerning the number of person- or animal-years at risk. As discussed previously, treating these count variables as continuous is not a big problem as long as their means are not close to zero. However, when the diseases are either non-contagious or rare, the mean of the observations could be very close to zero, and a more reasonable assumption in the analysis is that these counts follow some discrete distributions, such as Poisson distribution and negative binomial distribution.

10.4.1 Poisson Regression

In the following example, we consider the Poisson regression, which is a regression technique appropriate for Poisson-distributed data. To illustrate the usage of Poisson regression, we consider the daily COVID-19 death count for Los Angeles County, CA, and assume these county-level death counts (Y_i) follow a Poisson distribution.

In Poisson regression models, response variables are counts, such as the number of positive cases and deaths in a period. We assume that the conditional mean value of the mean of the response variable (μ) can be determined by a set of d predictors $(X_1, \ldots, X_d)$ via a $\log(\cdot)$ function as follows:

$$\log(\mu) = \beta_0 + \beta_1 X_1 + \cdots + \beta_d X_d. \tag{10.7}$$

In infectious disease studies, the terms $(\beta_0 + \beta_1 X_1 + \ldots + \beta_d X_d)$ in (10.7) can be used to represent an adjustment to account for disease counts that are either above or below that expected, based on the time at risk.

Similar to linear regression, Poisson regression requires model assumptions: (i) the response variable is a count per unit of time or space, described by a Poisson distribution; (ii) the observations must be independent of one another; (iii) the mean of a Poisson random variable must be equal to its variance; and (iv) the log of the mean rate, $\log(\mu)$, must be a linear function of the independent variable.

Example 10.6. (COVID-19 Death Count Prediction via the Shared-county Exponential Predictor). We illustrate how to use model (10.3) to make predictions for time series LA.ts based on the Poisson regression method. Suppose we are interested in making predictions for the period December 5 to December 11, 2020, and we use a training set from the past week (November 25 to December 4, 2020). Model (10.3) will be fitted on the training set, and the fitted model is used to make predictions. Figure 10.7 shows the forecasts of the daily death count. We also treat the response variable as a continuous variable and fit a linear regression model for comparison. Based on Figure 10.7, we can find that both models yield very similar results, which confirms our previous discussion that the two models work similarly when the mean of the observations is away from zero.

```
# Set training data from November 25 to December 4
train <- LA.ts %>%
  filter_index("2020-11-25" ~ "2020-12-04")
n <- nrow(train)
h <- 7
x <- log(train$Y.Death[1:(n-1)] + 1)
mfit.poi <- glm(train$Y.Death[2:n] ~ x, family = "poisson")
mfit.gau <- glm(train$Y.Death[2:n] ~ x, family = "gaussian")
Y.pred.poi <- rep(NA, h)
Y.pred.gau <- rep(NA, h)

for(j in 1:h){
  if(j == 1){
```

```
    x.new.poi <- data.frame(x = log(train$Y.Death[n] + 1))
    x.new.gau <- data.frame(x = log(train$Y.Death[n] + 1))
  }else{
    x.new.poi <- data.frame(x = log(Y.pred.poi[j-1] + 1))
    x.new.gau <- data.frame(x = log(Y.pred.gau[j-1] + 1))
  }
  Y.pred.poi[j] <- round(predict(mfit.poi, newdata = x.new.poi,
                                 type = "response"))
  Y.pred.gau[j] <- round(predict(mfit.gau, newdata = x.new.gau,
                                 type = "response"))
}

death.pred <- data.frame(DATE = train$DATE[n] + (1:h),
                         Y.Death.poi = Y.pred.poi,
                         Y.Death.gau = Y.pred.gau)



# Plot of reported vs predicted death count
ggplot() +
  geom_line(data = LA.ts %>%
                dplyr::filter(DATE < train$DATE[n] + h),
            aes(x = DATE, y = Y.Death, color = "Reported")) +
  geom_line(data = death.pred, aes(x = DATE, y = Y.Death.poi,
                                   color = "Predicted.poi")) +
  geom_line(data = death.pred, aes(x = DATE, y = Y.Death.gau,
                                   color = "Predicted.gau")) +
  scale_color_manual(
    values = c(Reported = "black", Predicted.gau = "cyan3",
               Predicted.poi = "red")) +
    labs(y = "Death count") +
    guides(color = guide_legend(title = "Series")) +
  theme(legend.position = "bottom")
```

10.4.2 Zero-inflated Poisson Regression

In the early stage of an epidemic, data quality is usually questionable with many issues, including inconsistent detection of cases, delayed reporting, and poor documentation. The restricted data quality will affect the quality of any model output. As a result, many counties have zero daily infected cases or deaths at the early stage of disease spread; for example, the reported death count for Los Angeles County, CA in Figure 10.7. To allow for frequent zero-

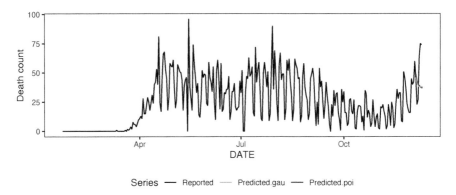

FIGURE 10.7: Reported and predicted death count for Los Angeles County via Poisson regression and ordinary linear regression.

valued observations, we usually consider a zero-inflated model based on a zero-inflated probability distribution in statistics. In the following example, we assume the observed counts Y contribute to a zero-inflated Poisson (ZIP) distribution, $\text{ZIP}(\mu, p)$. Expressly, we assume that

$$P(Y = y | X_1, \ldots, X_d) = \begin{cases} 1 - p, & y = 0, \\ p \frac{(\mu)^y}{\{\exp(\mu) - 1\} y!}, & y > 0, \end{cases}$$

where $p = \exp(\eta) / \{1 + \exp(\eta)\}$ and $\eta = a_1 + \exp(a_2) \log(\mu)$. Here, μ is generated from (10.7), and a_1, a_2 are unknown parameters to be estimated along with the roughness parameters. Detailed procedures for the estimation of a_1 and a_2 can be found in Wood et al. (2016).

To implement the zero-inflated Poisson regression, we can use the R package "mgcv." To compare the Poisson regression and zero-inflated Poisson regression, we consider the training data from two time periods: (1) the early stage of the pandemic (February 15 to April 15, 2020); and (ii) the later phase of the pandemic (November 25 to December 4, 2020). In the following, we work on the first training period. Based on Figure 10.8, we can see that the performances of the Poisson regression and the zero-inflated Poisson regression are entirely different. Without handling the zeros at the beginning of the pandemic, the Poisson regression predicts that deaths will increase rapidly in the following week. On the other hand, the zero-inflated Poisson regression predicts a more stable death count for next week.

```
# Set first training data from February 15 to April 15
train1 <- LA.ts %>%
    filter_index("2020-02-15" ~ "2020-04-15")
```

```
n <- nrow(train1)
h <- 7
x <- log(train1$Y.Death[1:(n-1)] + 1)
mfit.poi <- glm(train1$Y.Death[2:n] ~ x, family = "poisson")
mfit.zip <- gam(train1$Y.Death[2:n] ~ x, family = "ziP")
Y.pred.poi <- rep(NA, h)
Y.pred.zip <- rep(NA, h)

for(j in 1:h){
  if(j == 1){
    x.new.poi <- data.frame(x = log(train1$Y.Death[n] + 1))
    x.new.zip <- data.frame(x = log(train1$Y.Death[n] + 1))
  }else{
    x.new.poi <- data.frame(x = log(Y.pred.poi[j-1] + 1))
    x.new.zip <- data.frame(x = log(Y.pred.zip[j-1] + 1))
  }
  Y.pred.poi[j] <- round(predict(mfit.poi, newdata = x.new.poi,
                             type = "response"))
  Y.pred.zip[j] <- round(predict(mfit.zip, newdata = x.new.zip,
                             type = "response"))
}

death.pred <- data.frame(DATE = train1$DATE[n] + (1:h),
                       Y.Death.poi = Y.pred.poi,
                       Y.Death.zip = Y.pred.zip)
```

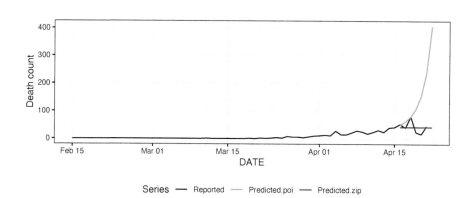

FIGURE 10.8: Reported and predicted death count for Los Angeles County via Poisson regression and zero-inflated Poisson regression; prediction is based on training data from February 15 to April 15, 2020.

Next, we consider the second training period (November 25 to December 4, 2020). We can see from Figure 10.9 that the predictions from Poisson regression and zero-inflated Poisson are identical.

```r
# Set second training data from November 25 to December 4
train2 <- LA.ts %>%
  filter_index("2020-11-25" ~ "2020-12-04")
n <- nrow(train2)
h <- 7
x <- log(train2$Y.Death[1:(n-1)] + 1)
mfit.poi <- glm(train2$Y.Death[2:n] ~ x, family = "poisson")
mfit.zip <- gam(train2$Y.Death[2:n] ~ x, family = "ziP")
Y.pred.poi <- rep(NA, h)
Y.pred.zip <- rep(NA, h)

for(j in 1:h){
  if(j == 1){
    x.new.poi <- data.frame(x = log(train2$Y.Death[n] + 1))
    x.new.zip <- data.frame(x = log(train2$Y.Death[n] + 1))
  }else{
    x.new.poi <- data.frame(x = log(Y.pred.poi[j-1] + 1))
    x.new.zip <- data.frame(x = log(Y.pred.zip[j-1] + 1))
  }
  Y.pred.poi[j] <- round(predict(mfit.poi, newdata = x.new.poi,
                          type = "response"))
  Y.pred.zip[j] <- round(predict(mfit.zip, newdata = x.new.zip,
                          type = "response"))
}

death.pred <- data.frame(DATE = train2$DATE[n] + (1:h),
                         Y.Death.poi = Y.pred.poi,
                         Y.Death.zip = Y.pred.zip)
```

10.4.3   Count Time Series Analysis

Following the ideas of generalized linear models, we can consider the analysis of count time series using the R package "tscount." The main function ts-glm() takes a "ts" object as an argument, with the ability to regress the conditional mean of counts on nonnegative past means and past observations, based

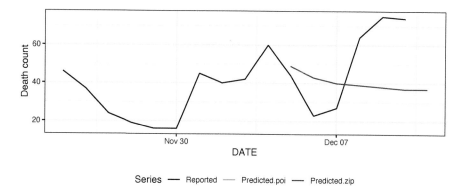

FIGURE 10.9: Reported and predicted death count for Los Angeles County via Poisson regression and zero-inflated Poisson regression; prediction is based on training data from November 25 to December 4, 2020. Here the two predictions overlap each other.

on the integer-valued generalized autoregressive conditional heteroscedasticity (INGARCH) model; see Ferland et al. (2006). In addition, the package allows identity and log link functions and supports Poisson and Negative Binomial as conditional distributions.

```
install.packages("tscount")
```

Example 10.7. (Simulation study comparing "glm" and "tscount"). First, we consider a simulation study, where Y_t, $t = 2, \ldots, 1000$ follows a Poisson distribution, and the logarithm of conditional mean has a linear form $\beta_0 + \beta_1 Y_{t-1}$, and the starting value $Y_1 = 10$. This autoregressive structure can be handled by specifying the indices of past observations that need to be included in the model. In this example, we specify model = list(past_obs = 1). The fitted coefficients are fairly close to the true coefficients $\beta_0 = 1.4$, $\beta_1 = 0.6$. We also conduct a Poisson regression using the glm() function for comparison. These two methods give very close estimators of the coefficients.

```
library(tscount)
# Simulation with preassigned parameters
nsim <- 1000
ysim <- c(10, rep(NA, nsim - 1))
beta0 <- 1.4
beta1 <- 0.6
```

```
for(i in 2:nsim){
  ysim[i] = rpois(n = 1,
                  lambda = exp(beta0 + beta1 * log(ysim[i-1] + 1)))
}

# Fit model using tsglm and retrieve coefficients
ts.fit <- tsglm(ts = ysim, link = "log",
                model = list(past_obs = 1), distr = "poisson")
ts.fit$coefficients
```

```
## (Intercept)       beta_1
##      1.4744       0.5807
```

```
# Fit model using glm and retrieve coefficients
m.fit <- glm(ysim[-1] ~ log(ysim[-nsim] + 1), family = "poisson")
m.fit$coefficients
```

```
##          (Intercept) log(ysim[-nsim] + 1)
##               1.4770               0.5802
```

Secondly, we consider the same setting as in Example 10.6, modeling the counts using the Poisson distribution with a log link. Here we see the easy predicting process when the model only involves auto-regressive structure, that we only need to specify the number of steps ahead, n.ahead, in the predict() function.

```
library(tscount)
# Set training data from November 25 to December 4, 2020
train <- LA.ts %>%
  filter_index("2020-11-25" ~ "2020-12-04")
n <- nrow(train)
h <- 7
ts.fit <- tsglm(ts = train$Y.Death, link = "log",
                model = list(past_obs = 1), distr = "poisson")
Y.pred <- round(predict(ts.fit, n.ahead = h)$pred)
death.pred.ts <- data.frame(DATE = train$DATE[n] + (1:h),
                            Y.Death = Y.pred)
```

Similarly, `tsglm()` also enables users to add external covariates in the argument as `xreg`. Below we demonstrate the case with 14-day lag infected counts as the external covariate.

```
# Add the infected count with 14-day lag as a covariate
xreg <- IDDA::I.county %>%
  dplyr::filter(State == 'California' & County == "LosAngeles")%>%
  dplyr::select(X2020.11.11:X2020.11.28) %>%
  pivot_longer(cols = X2020.11.11:X2020.11.28,
               names_to = "DATE",
               values_to = "Infected") %>%
  mutate(DATE = as.Date(DATE, "X%Y.%m.%d"))
ts.fit.X <- tsglm(ts = train$Y.Death, link = "log",
                  xreg = xreg$Infected[1:n],
                  model = list(past_obs = 1), distr = "poisson")
Y.pred.X <- round(predict(ts.fit.X, n.ahead = h,
                          newxreg = xreg$Infected[(n+1):(n+h)])$pred)
death.pred.ts.X <- data.frame(DATE = train$DATE[n] + (1:h),
                              Y.Death = Y.pred.X)
```

Then we compare the predictions by the R packages "glm," "tscount" without and with an external covariate. The zoomed-in plot in Figure 10.10 shows that the prediction produced by "glm" is between the two predictions produced by "tscount." The two predictions based on Poisson distribution, `Pred.poi` using "glm" and `Pred.tscount.poi` using "tscount," are not exactly the same due to the integer setup of "tscount."

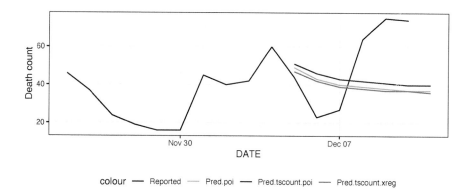

FIGURE 10.10: Reported and predicted death count for Los Angeles County via Poisson regression using `tsglm()` without and with an external covariate.

10.5   Logistic Regression

The spatial unit of interest typically shifts from areas to points due to the increasing level of the resolution of our analyses. Therefore, the objective usually switches to identifying factors that influence the risk of disease being present or absent at specific locations (e.g., farm or household) instead of describing and explaining disease counts summarized in different areas. Typically, we use the binary labels "positive" (disease present) or "negative" (disease absent) to describe the data.

When modeling binary data, explanatory variables are used to predict the probability of a study subject being disease present (a "case"). We can use the dummy variable to code the response

$$Y = \begin{cases} 0, & \text{if disease present;} \\ 1, & \text{if disease absent.} \end{cases}$$

Rather than modeling the response Y directly, logistic regression models the probability that Y belongs to a particular category. The logistic model solves the following problems:

$$\log \left\{ \frac{P(Y = 1|X = x)}{P(Y = 0|X = x)} \right\} = \beta_0 + \beta_1 x,$$

for some unknown β_0 and β_1, which we will estimate directly. Note that $P(Y = 0|X = x) = 1 - P(Y = 1|X = x)$, so we have

$$\log \left(\frac{p}{1 - p} \right) = \beta_0 + \beta_1 x,$$

where $p = P(Y = 1|X = x)$ is the probability that event Y occurs.

10.5.1   Odds and Odds Ratios

10.5.1.1   Odds

Odds are the probability of an event occurring (p) divided by the probability of the event not occurring ($1-p$), i.e., $odds = p/(1-p)$. For example, suppose that the probability of mortality is 0.25 in a group of patients. This can be expressed as the odds of dying: $0.25/(1 - 0.25) = 0.333$. The odds have a range from 0 to ∞. The values greater than one represent that an event is more likely to occur than not occur, while the values less than one mean that an event is less likely to occur than not.

The **logit** is defined as the log of the odds:

$$\log(\text{odds}) = \log\left(\frac{p}{1-p}\right) = \log(p) - \log(1-p).$$

Notice that the probabilities only range from 0 to 1. This transformation is functional because the variable will have a range from $-\infty$ to ∞ after the transformation. Therefore, it solves our problem of fitting a linear model to probabilities.

The interpretation of logits is very straightforward: it takes the exponential of the logit so that we can have the odds for the two groups.

10.5.1.2 Odds Ratio

For a continuous independent variable, the **odds ratio** can be defined as:

$$\frac{\text{odd}(x+1)}{\text{odd}(x)} = \frac{\exp\{\beta_0 + \beta_1(x+1)\}}{\exp(\beta_0 + \beta_1 x)} = \exp(\beta_1).$$

Thus, the **log odds ratio** is β_1. Although mathematically, the log odds ratios are easier to work with, it is more natural to use the odds ratios when it comes to interpretation. Unlike the log odds ratio, the odds ratio is always positive. Specifically, value one indicates no change, values between zero and one indicate a decreased probability of the outcome event, and values greater than one show an increased probability.

When the explanatory variable is categorical, the odds ratio represents the increase or decrease in odds over the baseline category. For a binary independent variable, the odds ratio is defined as the odds of an event happening under two different conditions. For example, consider the following contingency table:

	Disease	No Disease
Exposed	a	b
Unexposed	c	d

The odds ratio is defined as:

$$\text{OR} = \frac{a/c}{b/d} = \frac{ad}{bc}.$$

10.5.2 Estimating Logistic Regression Coefficients

Suppose that we are given a sample (x_i, y_i), $i = 1, \ldots, n$, where y_i denotes the class $\in \{0, 1\}$ of the ith observation. Assume that the classes are conditionally independent given $x_1, \dots, x_n$, then

$$\mathbb{L}(\beta_0, \beta_1) = \prod_{i=1}^{n} P(Y = y_i | X = x_i),$$

the likelihood of these n observations, so the log-likelihood is

$$l(\beta_0, \beta_1) = \sum_{i=1}^{n} \log P(Y = y_i | X = x_i).$$

For convenience, we define the indicator $u_i = I(y_i = 1)$, Then, the log-likelihood can be written as

$$
\begin{aligned}
l(\beta_0, \beta_1) &= \sum_{i=1}^{n} \log P(Y = y_i | x = x_i) \\
&= \sum_{i=1}^{n} \left[u_i(\beta_0 + \beta_1 x_i) - \log\{1 + \exp(\beta_0 + \beta_1 x_i)\} \right].
\end{aligned}
$$

The coefficients are estimated by maximizing the likelihood,

$$\sum_{i=1}^{n} \left[u_i(\beta_0 + \beta_1 x_i) - \log\{1 + \exp(\beta_0 + \beta_1 x_i)\} \right].$$

10.5.3 Logistic Regression with Multiple Explanatory Variables

Logistic regression is widely used to illustrate the exposure-response relationship. We can use it to assess the likelihood of infectious disease as a function of risk or exposure factors (and covariates). If there are multiple explanatory variables $X_1, \dots, X_p$, then we can consider the following logistic regression model

$$\log\left(\frac{p}{1-p}\right) = \beta_0 + \beta_1 X_1 + \dots + \beta_p X_p.$$

Odds ratios are the increase or decrease in odds associated with a unit change of the predictor with all other predictors being held constant.

The predicted value is

$$\hat{p} = \frac{e^{\hat{\beta}_0 + \hat{\beta}_1 X_1 + \dots + \hat{\beta}_p X_p}}{1 + e^{\hat{\beta}_0 + \hat{\beta}_1 X_1 + \dots + \hat{\beta}_p X_p}}.$$

To do a logistic regression analysis, we can implement the R function `glm()` using the `family = binomial` argument. We can use the `predict()` function along with the results of a `glm()` object to predict new data.

Example 10.8. (Logistic Regression for CDC FluView Portal Data).
For illustration, we consider the dataset `usflu.ts` from the CDC FluView
Portal Data. We would like to explain whether weekly infected counts between
age 0 and age 4 exceed 4,000 using time of year. We divide the data into
training and testing sets, with the latest 52-week observations as testing data
and others as training data. First, we define the binary random variable as
`age_0_4_higherthan_4000`, which takes a value of 1 if the weekly counts exceed
4,000, and 0 otherwise. Then, we consider the model

$$\log\left(\frac{p}{1-p}\right) = \beta_0 + \beta_1 \sin(2\pi t_{ij}) + \beta_2 \cos(2\pi t_{ij}),$$

where $\sin(2\pi t_{ij})$ and $\cos(2\pi t_{ij})$ are regarded as X_{1i} and X_{2i} respectively, p
represents the conditional probability $P(Y_i = 1 | X_1 = x_{1i}, X_2 = x_{2i})$.

```
ind_test = seq(nrow(usflu) - 51, nrow(usflu))
# Create a new binary variable
usflu$age_0_4_higherthan_4000 = ifelse(usflu$age_0_4 > 4000, 1, 0)
train = usflu[-ind_test,]
test = usflu[ind_test,]
fit_logi = glm(formula = age_0_4_higherthan_4000 ~ 1 +
                  sin(time_in_year * 2 * pi) +
                  cos(time_in_year * 2 * pi),
               data = train,
               family = binomial)
coef_logi = fit_logi$coefficients
usflu$logi_phat = NA
usflu$logi_preds = NA

# Record the predicted conditional probability
# and predict Y = 1 if p > 0.5
usflu$logi_phat[ind_test] =
  predict(fit_logi, newdata = test, type = "response")
usflu$logi_preds[ind_test] =
  ifelse(predict(fit_logi, newdata = test, type = "response") > 0.5,
         1, 0)

p2 <- ggplot(usflu, aes(x = week)) +
    geom_line(aes(y = age_0_4_higherthan_4000, color = "Reported")) +
    geom_line(aes(y = logi_phat, color = "Logistic_p")) +
    geom_line(aes(y = logi_preds, color = "Logistic")) +
    scale_color_manual(
      values = c(Reported = "black", Logistic_p = "cyan3",
                 Logistic = "red")) +
    labs(x = "week",
```

```
        title = "Reported vs logistic regression fit") +
  guides(color = guide_legend(title = "Series")) +
  theme(legend.position = "bottom")
```

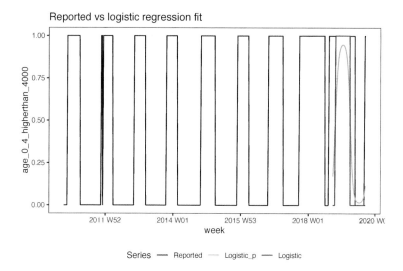

FIGURE 10.11: Whether national influenza-like illness weekly counts for age 0 - 4 exceed 4000 (black) vs predicted conditional probability (cyan) vs logistic regression prediction (red).

Notice the prediction of conditional probability $\hat{p}$ (in cyan) is a smooth function bounded between 0 and 1, in contrast to the observations (in black), which only take values at 0 and 1. This is due to the fact that in logistic regression, we model the probability of the occurrence of an event, i.e., $p = P(Y = 1|X = x)$, instead of the value of the binary response variable (Y). Therefore, we need to take one more step to get the prediction of Y by looking at whether $\hat{p} > 0.5$. If so, we say $\hat{Y} = 1$, otherwise $\hat{Y} = 0$. From Figure 10.11, we find the predicted response $\hat{Y}$ (in red) is able to capture the trend of Y but underestimates at the beginning and the end of the prediction period.

10.6 Further Reading

The following books provide many valuable resources with regard to the regression topics covered in this chapter:

- James, G., Witten, D., Hastie, T. and Tibshirani, R. (2013). *An introduction to statistical learning.* New York: Springer.
- Ruppert, D., Wand, M. P. and Carroll, R. J. (2003). *Semiparametric regression.* Cambridge University Press.

For the R packages introduced in this chapter, there are valuable resources available:

- `https://cran.r-project.org/web/packages/mgcv;`
- `https://cran.r-project.org/web/packages/tscount/index.html;`
- Wood, S. N. (2006). *Generalized additive models: An introduction with R.* Chapman and Hall/CRC.

10.7 Exercises

1. From the data `state.ts` in the R package IDDA, choose a state and choose any quarter of the data. Hold out the last seven days as test data and use all the data except the last seven days as your training data. Fit the following models based on your training data.

   (a) Piecewise constant spline regression model with 15 interior knots.
   (b) Truncated linear spline regression model with 10 interior knots.
   (c) Natural spline regression model with 8 interior knots.
   (d) Smoothing spline regression model with knots automatically selected by the "mgcv" package.

2. Demonstrate different model fits in Problem 1 in one time series plot. In addition, for each model fitted, make some residual plots, such as the ACF and the histogram. Provide some comments about the model fit and model comparison.

3. Use the fitted natural spline regression model in Problem 1 to make a 7-day-ahead prediction, and show the 95% prediction intervals.

4. Choose a county from `CA.county.ts` in the R package IDDA, and work on the time series of daily new deaths. Consider a two-week period and use it as training data to forecast the daily new death count in the following week. Try each of the following predictors:

   (a) the separate-county exponential predictor;
   (b) the separate-county linear predictor;
   (c) the expanded shared-county exponential predictor.

11

Neural Networks

Machine learning methods have gained increasing popularity in epidemiological modeling and projection due to their great flexibility to capture disease spread patterns. This chapter describes a popular machine learning method for epi-forecasting called neural networks. An artificial neural network is a computing system inspired by a biological neural network that constitutes the human brain, which learns about underlying relationships in data by considering examples and finding similarities. This chapter starts with an introduction to the building blocks of a neural network, and then it describes how a neural network can be trained. The chapter also presents a class of neural network autoregressive models to make predictions for time series data, which combines traditional statistical models and neural networks. The chapter closes by demonstrating the usage of neural network autoregressive models with an application to COVID-19 data.

11.1 A Single Neuron

The basic computation unit in a neural network is the **node** or **unit** patterned after a neuron in a human brain. It receives input from some other nodes or an external source, processes the information, and produces an output. Each input has an associated weight (β), which is assigned based on its relative importance to other inputs. As illustrated in Figure 11.1, the node applies a function g (see the definition below) to the weighted sum of its inputs.

The network in Figure 11.1 takes numerical inputs X_0, X_1, X_2, X_3 and has weights $\beta_0, \beta_1, \beta_2, \beta_3$ associated with those inputs. The output $a_1^{(2)}$ from the neuron in Figure 11.1 is computed using $g(\sum_{k=0}^{3} \beta_k X_k)$, where g introduces non-linearity into the output of a neuron, and is called the **activation function**. The activation functions are important in neural network analysis because most real data is nonlinear, and we want neurons to learn these nonlinear representations.

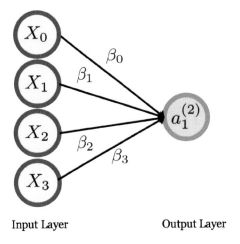

Input Layer   Output Layer

FIGURE 11.1: An illustration of a single neuron.

In a neural network, an activation function transforms the summed weighted input from the node into an output value to be fed to the next hidden layer or as output. There are several activation functions we may encounter in practice:

- **Sigmoid:** It takes a real-valued input and maps the resulting values in between 0 and 1

$$g(x) = 1/\{1 + \exp(-x)\}.$$

- **Tanh:** It takes a real-valued input and squashes it to the range $[-1, 1]$

$$g(x) = 2/\{1 + \exp(-2x)\} - 1.$$

- **ReLU (Rectified Linear Unit):** It takes a real-valued input and thresholds it at zero (replaces negative values with zero)

$$g(x) = \max(0, x).$$

Figure 11.2 shows each of the above activation functions.

11.2   Neural Network Structure

A neural network is made up of vertically stacked components called **layers**. Figure 11.3 shows a typical neural network constructed from three types of layers:

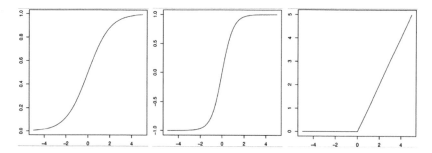

FIGURE 11.2: An illustration of three activation functions: Sigmoid (left panel), Tanh (middle panel), and ReLU (right panel).

Input layer: The input layer has four nodes with $X_0 = 1$. The other three nodes take X_1, X_2 and X_3 as external inputs (numerical values depending on the input data). The outputs from nodes in the input layer are X_0, X_1, X_2 and X_3, respectively, which are fed into the hidden layer.

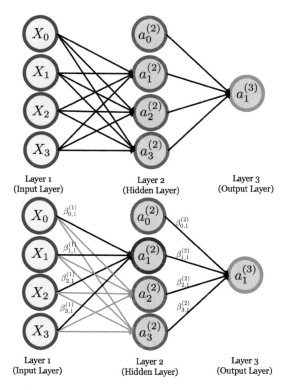

FIGURE 11.3: Top: a neural network with one hidden layer. Bottom: coefficients of a neural network with one hidden layer.

Hidden layer: The hidden layer also has four nodes with the bias node $a_0^{(2)}$. The output of the other two nodes in the hidden layer depends on the outputs from the input layer (X_0, X_1, X_2, X_3) as well as the weights associated with the connections (edges). For the hidden neuron i, let $a_i^{(2)}$ be its activation, then we have

$$a_i^{(2)} = g\left(\sum_{j=0}^{3} \beta_{j,i}^{(1)} X_j\right),$$

where g refers to the activation function. The bottom plot in Figure 11.3 shows the output calculation for the hidden node $a_1^{(2)}$.

Output layer: These outputs are then fed to the nodes in the output layer. In this simple example, the output layer has one node which takes inputs from the hidden layer. Similarly, the output $a_1^{(3)}$ can be calculated using

$$a_1^{(3)} = g\left(\sum_{j=0}^{3} \beta_{j,1}^{(2)} a_j^{(2)}\right).$$

Let's make it more general for a neural network with multiple hidden layers. For any $j = 1, \dots, L$, let $a_i^{(j)}$ be the activation of unit i in layer j, and let $\beta^{(j)}$ be a weight matrix that stores parameters from layer j to layer $j + 1$. If the network has s_j units in layer j and s_{j+1} units in layer $j + 1$, then $\beta^{(j)}$ has dimension $s_{j+1} \times (s_j + 1)$.

11.3   Neural Network Training

Neural network models have unknown parameters (weights). We can train a neural network using an optimization algorithm to find a set of parameters to best map inputs to outputs. For regression, we use the sum of squared errors as our measure of fit (error function). Similarly, our objective function in neural network training is

$$\ell(\beta^{(1)}, \dots, \beta^{(L)}) = \sum_{i=1}^{N} (Y_i - \hat{Y}_i)^2,$$

where L is the number of layers, and N is the number of observations in the training set.

Minimizing the above function $\ell(\cdot)$ is challenging since it is non-convex with the presence of local minima, flat regions, and the high-dimensionality of the search space. Typically we don't want the global minimizer of $\ell(\cdot)$, as this is likely to be an overfit solution. Instead, we can consider some regularization techniques, which can be achieved directly through a penalty term or indirectly by early stopping.

The generic approach to minimizing $\ell(\cdot)$ is by **gradient descent**, an optimization algorithm for finding a local minimum of a differentiable function. First, we start out by randomly guessing the parameters. Then, we evaluate the direction in which the loss function moves downward most steeply with respect to changing the parameters, and step slightly in that direction. This process will be repeated until we are satisfied with the lowest point that we have found.

To search for the direction in which the loss function moves downward most steeply, we need to calculate the gradient of the loss function with respect to all of the parameters. Using the chain rule for differentiation, we can easily derive the gradient because of the model's compositional form. To compute the gradients, we can utilize a forward and backward sweep over the network, keeping track only of quantities local to each unit. For any $j = 1, \dots, L$, a gradient descent has the form:

$$\beta^{(j)} \leftarrow \beta^{(j)} - \alpha \nabla_{\beta^{(j)}} \ell(\beta^{(1)}, \dots, \beta^{(L)}),$$

where $\nabla_{\beta^{(j)}} \ell(\beta^{(1)}, \dots, \beta^{(L)})$ is the partial derivative of $\ell(\beta^{(1)}, \dots, \beta^{(L)})$ with respect to $\beta^{(j)}$, and α is the **learning rate**. When α is too big, gradient descent may jump across the valley and end up on the other side. This will lead to the error function divergence. When α is too small, it will take the algorithm a long time to converge. Therefore, we need to have a proper learning rate before starting the gradient descent.

A neural network training algorithm finds the parameters with the help of forward propagation and backpropagation.

11.3.1 Forward Propagation

We start with one single observation point $(Y_i, \mathbf{X}_i)$. As shown in Figure 11.4, we can use the following equations to form the network's forward propagation.

$$
\begin{aligned}
\mathbf{a}^{(1)} &= \mathbf{X}_i, \\
\mathbf{z}^{(2)} &= \beta^{(1)\top} \mathbf{a}^{(1)}, \\
\mathbf{a}^{(2)} &= g(\mathbf{z}^{(2)}), \ (\text{add } a_0^{(2)}), \\
\mathbf{z}^{(3)} &= \beta^{(2)\top} \mathbf{a}^{(2)}, \\
\mathbf{a}^{(3)} &= g(\mathbf{z}^{(3)}), \ (\text{add } a_0^{(3)}), \\
\mathbf{z}^{(4)} &= \beta^{(3)\top} \mathbf{a}^{(3)}, \\
\mathbf{a}^{(4)} &= g(\mathbf{z}^{(4)}).
\end{aligned}
$$

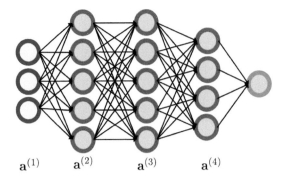

$$\mathbf{a}^{(1)} \qquad \mathbf{a}^{(2)} \qquad \mathbf{a}^{(3)} \qquad \mathbf{a}^{(4)}$$

FIGURE 11.4: A neural network with three hidden layers.

The final step in a forward pass is to evaluate the predicted output $\hat{Y}_i$ against an observed output Y_i. Evaluation between $\hat{Y}_i$ and Y_i happens through an error function, which can be the mean squared error in the regression case.

11.3.2 Backpropagation

Backpropagation is an efficient method to compute gradients needed to perform gradient-based optimization of the weights in a neural network.

11.3.2.1 Backpropagation: Top Layer

To minimize the error function ℓ, we want to find partial derivatives:

$$\frac{\partial \ell}{\partial \beta_{01}^{(2)}}, \ldots, \frac{\partial \ell}{\partial \beta_{31}^{(2)}},$$

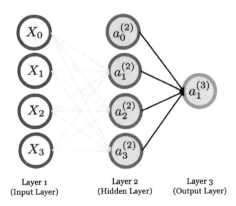

FIGURE 11.5: A neural network with one hidden layer.

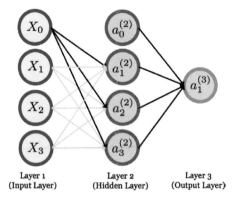

FIGURE 11.6: A neural network with one hidden layer.

and we can show, for example,

$$\frac{\partial \ell}{\partial \beta_{01}^{(2)}} = \frac{\partial \ell}{\partial a_1^{(3)}} \frac{\partial a_1^{(3)}}{\partial z_1^{(3)}} \frac{\partial z_1^{(3)}}{\partial \beta_{01}^{(2)}}$$

$$= 2(Y - a_1^{(3)}) g'(z_1^{(3)}) a_0^{(2)},$$

where g' is the derivative of the activation function g in the output layer.
Then, we update $\beta_{0,1}^{(2)}$ by

$$\beta_{0,1}^{(2)} \leftarrow \beta_{0,1}^{(2)} - \alpha \frac{\partial \ell}{\partial \beta_{0,1}^{(2)}}.$$

11.3.2.2 Backpropagation: Next Layer

Next, we want to find the following partial derivatives:

$$\frac{\partial \ell}{\partial \beta_{0,1}^{(1)}}, \cdots, \frac{\partial \ell}{\partial \beta_{0,3}^{(1)}},$$

and for example, we have

$$\frac{\partial \ell}{\partial \beta_{01}^{(2)}} = \frac{\partial \ell}{\partial a_1^{(3)}} \frac{\partial a_1^{(3)}}{\partial z_1^{(3)}} \frac{\partial z_1^{(3)}}{\partial \partial a_1^{(2)}} \frac{\partial a_1^{(2)}}{\partial z_1^{(2)}} \frac{\partial z_1^{(2)}}{\partial \beta_{01}^{(2)}}$$

$$= 2(Y - a_1^{(3)}) g'(z_1^{(3)}) \beta_{11}^{(2)} g'(z_1^{(2)}) X_0,$$

where g' is the derivative of the activation function in the output layer and g'
is the derivative of the activation function in the second layer.

Then, we update $\beta_{01}^{(1)}$ by

$$\beta_{0,1}^{(1)} \leftarrow \beta_{0,1}^{(1)} - \alpha \frac{\partial \ell}{\partial \beta_{0,1}^{(1)}}.$$

Initial values of the weights and biases β are randomly chosen. The derivative of ℓ can be calculated using the partial derivatives described above. The termination condition is met once the error function is minimized.

11.3.2.3 Backpropagation: Jacobian Matrix

The Jacobian matrix of a vector-valued function of several variables is the matrix of all its first-order partial derivatives. Suppose $f : R^n \rightarrow R^m$, the Jacobian matrix of function f is

$$J = \frac{\partial f}{\partial x} = \begin{bmatrix} \frac{\partial f_1}{\partial x_1} & \cdots & \frac{\partial f_1}{\partial x_n} \\ \vdots & \ddots & \vdots \\ \frac{\partial f_m}{\partial x_1} & \cdots & \frac{\partial f_m}{\partial x_n} \end{bmatrix}.$$

We have the following chain rule:

$$\frac{\partial f\{g(x)\}}{\partial x} = \frac{\partial f\{g(x)\}}{\partial g(x)} \times \frac{\partial g(x)}{\partial x}.$$

11.4 Overfitting

Often neural networks have too many weights and will overfit the data. A solution to overcome this overfitting issue is to update the training algorithm and encourage the network to keep the weights small. This is called **weight regularization**, which penalizes the network's large weights. Specifically, we add a penalty to the error function,

$$\sum_{j=1}^{L} \|\beta^{(j)}\|_F, \quad \|\beta^{(j)}\|_F = \sum_{k=1}^{s_j} \sum_{k'=1}^{s_{j+1}} (\beta_{k,k'}^{(j)})^2,$$

where $\|\beta^{(j)}\|_F$ is the Frobenius norm of matrix $\beta^{(j)}$. We minimize the following objective function

$$\ell(\beta^{(1)}, ..., \beta^{(L)}) + \rho \sum_{j=1}^{L} \|\beta^{(j)}\|_F, \tag{11.1}$$

where ρ is the penalization parameter.

Dropout is another regularization strategy that prevents neural networks from overfitting. While the above weight regularization reduces overfitting by modifying the loss function, dropouts deactivate a certain number of neurons at a layer from firing during training. This technique is called **dropout regularization**, which mitigates the overfitting problem by preventing the network from relying too much on single neurons and forces all neurons to learn to generalize better. We refer readers to Srivastava et al. (2014) for more details of dropout regularization.

The R package "neuralnet" can be used to train and test a neural network. See more details on the CRAN webpage[1].

11.5 Neural Network Auto-Regressive (NNAR) Models

To bridge the statistical and machine learning-based approaches in time series analysis, Tang and Fishwick (1993) introduced a class of feed-forward neural networks for time series forecasting, in which lagged values of the time series are used as inputs to a neural network.

When training the neural network for time series data, one of our objectives is to determine how many lags to include in the input layer and how many neurons to include in the hidden layer to maximize the forecasting accuracy. There are no fixed rules on the number of layers to use or the number of nodes to use in each layer. This section only considers feed-forward networks with one hidden layer. Let us use $\text{NNAR}(p, k)$ to denote a single hidden layer neural network with p lagged inputs and k nodes in the hidden layer. For example, an NNAR(9,5) model is a neural network with the last nine observations $(y_{t-1}, y_{t-2}, \dots, y_{t-9})$ used as inputs for forecasting the output y_t, and with five neurons in the hidden layer.

As shown in Hyndman and Athanasopoulos (2018), an $\text{NNAR}(p, 0)$ model is equivalent to an $\text{ARIMA}(p, 0, 0)$ model, but without the restrictions on the parameters to ensure stationarity.

For seasonal data, we can consider seasonal $\text{NNAR}(p, P, k)_m$ with period m, which uses inputs $(Y_{t-1}, Y_{t-2}, \dots, Y_{t-p}, Y_{t-m}, Y_{t-2m}, \dots, Y_{t-Pm})$ and k neurons in the hidden layer. For example, an $\text{NNAR}(3, 1, 2)_7$ model has inputs $(Y_{t-1}, Y_{t-2}, Y_{t-3}, Y_{t-7})$, and two neurons in the hidden layer. An $\text{NNAR}(p, P, 0)_m$ model is equivalent to an $\text{SARIMA}(p, 0, 0) \times (P, 0, 0)_m$ model but without stationarity restrictions.

---

[1] https://cran.r-project.org/web/packages/neuralnet/

The nnetar() function in the "forecast" R package fits an NNAR$(p, P, k)_m$ model. If p and P are not specified, they are automatically selected. For non-seasonal time series, default p is set to be the optimal number of lags for a linear AR(p) model according to the AIC. For seasonal time series, by default, $P = 1$ and p is chosen from the optimal linear model fitted to the seasonally adjusted data.

The network is applied iteratively to make a forecast. For a one-step-ahead forecast, we simply use the available historical inputs. To make a two-step-ahead forecast, we use the one-step forecast as an input, along with the historical data. The process proceeds until we have computed all the required forecasts.

11.6 COVID-19 Forecasting Using NNAR

In this section, we illustrate how to apply the NNAR approach to COVID-19 forecasting. To forecast the future number of cases, we use its w lags, and define $\mathbf{y} = (y_{n-w+1}, \ldots, y_{n-1}, y_n)^\top \in R^w$. Suppose we are interested in predicting $y_{n+h} \in R$ based on the historical data $\mathbf{y}$. We can use cross-validation to find the proper window size w, the number of hidden neurons L, and the regularization hyperparameter ρ in (11.1).

The nnetar() function in the "forecast" R package fits a feed-forward neural networks model with a single hidden layer and lagged inputs for forecasting univariate time series. It uses lagged values of the time series as inputs (and possibly some other exogenous inputs).

When using the "forecast" package, it is better for us to store the time series as a ts object in R. Let us consider the time series of daily new deaths for Florida, which can be obtained from the state.ts dataset in the IDDA package.

```
# devtools::install_github('FIRST-Data-Lab/IDDA')
library(tidyverse)
library(IDDA)
data(state.ts)

Florida.ts <- state.ts %>%
    dplyr::filter(State == "Florida")
```

We first turn Florida.ts into a ts object using the ts() function. There is a frequency argument in the ts() function, which needs to be set appropriately for the data. Here frequency is the number of observations per "cycle." For annual data, frequency = 1; for quarterly data, frequency = 4; for monthly data, frequency = 12, and for weekly data, frequency = 52. There are multiple ways of handling the frequency if the frequency of observations is greater than once per week. For example, data with daily observations might have a weekly seasonality (frequency = 7) or an annual seasonality (frequency =365.25). Read the online materials on Seasonal Periods[2] for an explanation about how to set the frequency for the seasonal period in ts objects. Since we have observed the 7-day cycle for this data, we can simply add a frequency = 7 argument.

```
y <- ts(Florida.ts$Y.Death, start = c(2020, 1), frequency = 7)
```

Simply applying the default nnetar() function, we obtain the feed-forward neural networks model $NNAR(21, 1, 11)_7$. It is based on an average of 20 networks, each of which is $NNAR(21, 1, 11)_7$ with 254 weights.

```
library(forecast)
set.seed(2020)
fit1 <- nnetar(y)
fit1
```

```
## Series: y
## Model:   NNAR(21,1,11)[7]
## Call:    nnetar(y = y)
##
## Average of 20 networks, each of which is
## a 21-11-1 network with 254 weights
## options were - linear output units
##
## sigma^2 estimated as 8.67
```

```
nnar.fc.report1 <- as.data.frame(forecast(fit1, h = 14)) %>%
  mutate(DATE = as.Date("2020-12-31") + 1:14)
names(nnar.fc.report1) <- c("Y.Death", "DATE")
```

The top panel in Figure 11.7 shows the two-week-ahead forecast of the daily new deaths for Florida based on $NNAR(21, 1, 11)_7$.

---

[2]https://robjhyndman.com/hyndsight/seasonal-periods/

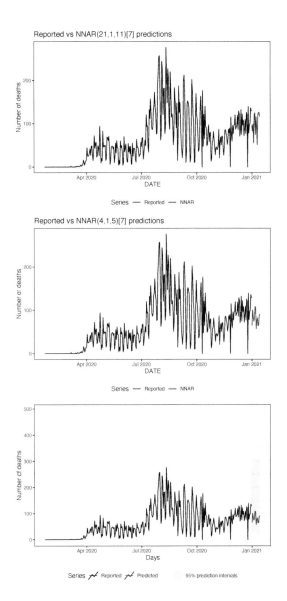

FIGURE 11.7: Daily new deaths for Florida. Top: two-week-ahead forecast using nnetar. Center: two-week-ahead forecast using nnetar with predictions constrained to positive values. Bottom: two-week-ahead forecast and prediction intervals using nnetar with predictions constrained to positive values.

```
# Plot of reported vs NNAR predictions
 ggplot() +
    geom_line(aes(x = DATE, y = Y.Death, color = "Reported"),
             data = Florida.ts) +
    geom_line(aes(x = DATE, y = Y.Death, color = "NNAR"),
             data = nnar.fc.report1) +
    scale_color_manual(
      values = c(Reported = "black", NNAR = "red")
    ) +
    labs(y = "Number of deaths",
        title = "Reported vs NNAR(21,1,11)[7] predictions") +
    guides(color = guide_legend(title = "Series"))+
    theme(legend.position = "bottom")
```

We can also specify the `repeats` and `size` options in the `nnetar` function. The `repeats` option of `nnetar` shows the number of networks (default = 20) to fit with different random starting weights. These networks are then averaged when producing forecasts. The `size` option provides the number of nodes in the hidden layer. The default is half of the number of input nodes (including external regressors, if given) plus 1.

We can consider a Box-Cox transformation for the data. For example, if we want to restrict to positive values, we can consider a Box-Cox transformation parameter `lambda`. If `lambda = 0`, then a natural logarithm transformation is applied, which forces forecasts to be positive. If `lambda = auto`, then a transformation is automatically selected using `BoxCox.lambda`. The transformation is ignored if `NULL`. Otherwise, data is transformed before the model is estimated.

The following code with `lambda = 0` and `size = 5` yields a $NNAR(4, 1, 5)_7$. The middle panel in Figure 11.7 shows its two-week-ahead forecast of the daily new deaths for Florida, where the predictions are constrained to positive values.

```
fit2 <- nnetar(y + 1, lambda = 0, size = 5)
fit2
```

```
## Series: y + 1
## Model:   NNAR(4,1,5)[7]
## Call:    nnetar(y = y + 1, size = 5, lambda = 0)
##
## Average of 20 networks, each of which is
## a 5-5-1 network with 36 weights
```

```
## options were - linear output units
##
## sigma^2 estimated as 0.346
```

```
nnar.fc.report2 <- as.data.frame(forecast(fit2, h = 14)) %>%
  mutate(DATE = as.Date("2020-12-31") + 1:14)
names(nnar.fc.report2) <- c("Y.Death", "DATE")
```

```
# Plot of reported vs NNAR predictions
 ggplot() +
     geom_line(aes(x = DATE, y = Y.Death, color = "Reported"),
                 data = Florida.ts) +
     geom_line(aes(x = DATE, y = Y.Death, color = "NNAR"),
                 data = nnar.fc.report2) +
     scale_color_manual(
       values = c(Reported = "black", NNAR = "red")
     ) +
     labs(y = "Number of deaths",
           title = "Reported vs NNAR(4,1,5)[7] predictions") +
     guides(color = guide_legend(title = "Series")) +
     theme(legend.position = "bottom")
```

Note that the NNAR model is a nonlinear autoregressive model, and it is not possible to analytically derive prediction intervals. Below we use a bootstrap method to generate simulated paths. Consider the following NNAR(p, k) model:

$$Y_t = f(Y_{t-1}, \dots, Y_{t-p}) + \varepsilon_t,$$

where f is a neural network with k hidden nodes in a single layer. The error series $\{\varepsilon_t\}$ is assumed to be homogeneous.

We can simulate future sample paths of this model iteratively, by randomly generating a value for ε_t, either from a normal distribution or by resampling from the historical values. For example, let ε^*_{T+1} be a random draw from the distribution of errors at time $T + 1$, then we can obtain

$$Y^*_{T+1} = f(Y_T, \dots, Y_{T-p+1}) + \varepsilon^*_{T+1},$$

which is one possible draw from the forecast distribution for Y_{T+1}. We can then repeat the process to get

$$Y^*_{T+2} = f(Y^*_{T+1}, Y_T, \dots, Y_{T-p+1}) + \varepsilon^*_{T+2},$$

and obtain a sample path $\{Y^*_{T+1}, Y^*_{T+2}, \dots, Y^*_{T+h}\}$.

If we generate many sample paths, we can get a good picture of the forecast distributions, which can be used to construct the prediction intervals. For example, as shown below, we set up npaths = 500 to generate 500 simulation bootstrap paths and construct prediction intervals.

```
fcast <- forecast(fit2, PI = TRUE, h = 14, npaths = 500)
#autoplot(fcast, xlab = "Days", ylab = "Daily new deaths")
fcast.report <- as.data.frame(fcast) %>%
  mutate(DATE = as.Date("2020-12-31") + 1:14)
names(fcast.report) <- c("Y.Death", "Lo.80", "Hi.80",
                         "Lo.95", "Hi.95", "DATE")
```

```
ggplot(Florida.ts, aes(DATE, Y.Death)) +
    geom_line(aes(color = 'Reported')) +
    labs(x = "Days", y = "Number of deaths") +
    # Add prediction intervals
    geom_ribbon(mapping = aes(x = DATE,
                              y = Y.Death,
                              ymin = Lo.95,
                              ymax = Hi.95,
                              fill = '95% prediction intervals'),
                data = fcast.report, alpha = 0.4) +
    # Add a line for predicted values
    geom_line(mapping = aes(x = DATE,
                            y = Y.Death,
                            color = 'Predicted'),
              data = fcast.report,
              key_glyph = "timeseries") +
    scale_color_manual(values = c(Reported = "black",
                                  Predicted = "red")) +
    scale_fill_manual("", values = "pink") +
    theme(legend.position = "bottom")
```

Since prediction intervals are calculated through simulations, the process can be slow. In the forecast.nnetar() function, by default, PI = FALSE, so prediction intervals are not computed unless requested. The npaths argument controls how many simulations to be conducted. By default, the errors are drawn from a normal distribution. Using the bootstrap argument, we can generate the errors randomly from the historical errors. The bottom panel in Figure 11.7 shows the prediction intervals based on $NNAR(4, 1, 5)_7$ with predictions constrained to positive values.

11.7   Further Reading

The following resources are helpful with regard to the topic of neural networks covered in this chapter.

- Hastie, T., Tibshirani, R., Friedman, J. H. and Friedman, J. H. (2009). *The elements of statistical learning: Data mining, inference, and prediction.* New York: Springer.
- Aguila, M. D. M. R. D., Requena, I., Bernier, J. L., Ros, E. and Mota, S. (2004). Neural networks and statistics: A review of the literature. *Soft Methodology and Random Information Systems*, 597-604.

For readers interested in learning how to train a neural network using the R "neuralnet" package, the following online resources are helpful:

- `https://cran.r-project.org/web/packages/neuralnet`;
- `https://datascienceplus.com/neuralnet-train-and-test-neural-networks-using-r/`;
- `https://www.datacamp.com/community/tutorials/neural-network-models-r`.

For a good reference to the NNAR models covered in this chapter, see:

- Hyndman, R.J. and Athanasopoulos, G. (2021). *Forecasting: Principles and practice.* 3rd edition, OTexts: Melbourne, Australia. Available at `https://otexts.com/fpp3/`.

11.8   Exercises

1. Based on the input values and weights given in Figure 11.8, find the values of $a_1^{(2)}$, $a_2^{(2)}$ and $a_1^{(3)}$. Use the sigmoid activation function.

2. From `state.long` in the IDDA package, choose a state and focus on the daily new deaths.

   (a) Apply the `nnetar` to your time series directly and report the fitted model.

   (b) Based on the model in part (a), make a 7-day-ahead forecast, and provide the 95% prediction intervals.

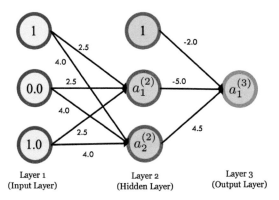

FIGURE 11.8: An example of neural network with one hidden layer.

(c) Apply the nnetar to the log-transformed time series and report the fitted model.

(d) Based on the model in part (c), make a 7-day-ahead forecast, and provide the 95% prediction intervals.

12

Hybrid Models

Although mechanistic and phenomenological models are two of the main types of models in epidemic modeling, in COVID-19 studies, an increasing number of hybrid methods have been developed. Typically, part of a hybrid model is formulated based on mechanistic principles, while part of the model needs to be inferred from data due to the lack of understanding of the details of the mechanistic. In previous chapters, we introduced compartment models, time series models, and some machine learning methods. This chapter mainly focuses on ensembling various models to make hybrid models. It introduces two R packages: "forecastHybrid" and "opera" for forecast combinations, and dives into specific examples to illustrate the usage. This chapter cannot be comprehensive with regard to hybrid forecasting methods, but it provides a number of examples of the popular ensemble forecasting tools used in data science.

12.1   Ensembling Time Series Models

As one of the widely used hybrid methods in forecasting, an **ensemble method** usually combines multiple algorithms so that it can improve the predictive performance of a simple prediction algorithm and avoid possible overfit. The basic idea of the ensemble method is that when there is much uncertainty in finding the best model in practice, combining multiply models may reduce the forecast instability and thus improve prediction accuracy.

As illustrated in Figure 12.1, a linear combination technique calculates the combined forecast for the variable of interest based on a linear function of the individual forecasts given by multiple contributing models.

Let $\{y_1, y_2, \dots, y_n\}$ be the actual time series. We are interested in making an h-step-ahead forecast using K different individual models. Let $\{\widehat{y}_{n+1}^{(k)}, \widehat{y}_{n+2}^{(k)}, \dots, \widehat{y}_{n+h}^{(k)}\}$ be its forecast obtained from the kth model ($k = 1, 2, \dots, K$). Then, a combination of these K forecasted series of the original

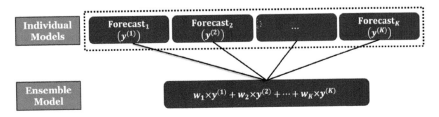

FIGURE 12.1: An illustration of ensemble methods.

time series produces $\{\widehat{y}^*_{n+1}, \widehat{y}^*_{n+2}, \ldots, \widehat{y}^*_{n+h}\}$, where

$$\widehat{y}^*_{n+j} = f\left(\widehat{y}^{(1)}_{n+j}, \ldots, \widehat{y}^{(K)}_{n+j}\right)$$

for $j = 1, 2, \ldots, h$. If f is some linear function, then we have

$$\widehat{y}^*_{n+j} = w_1 \widehat{y}^{(1)}_{n+j} + w_2 \widehat{y}^{(2)}_{n+j} + \cdots + w_K \widehat{y}^{(K)}_{n+j} = \sum_{k=1}^{K} w_k \widehat{y}^{(k)}_{n+j},$$

for $j = 1, 2, \ldots, h$. Here, w_k is the weight assigned to the kth forecasting method. Without loss of generality, the weights are assumed to sum to unity. We describe some widely used linear combination-based ensembling techniques as follows:

- Equal-weight average: assigns equal weights to all models; that is, for $k = 1, 2, \ldots, K$, we set $\omega_k = 1/K$.
- Trimmed average: combines individual forecasts by simple arithmetic means, excluding $\alpha\%$ of the models which perform worst, and α is often suggested to be taken from 10 to 30.
- Winsorized average: selects the m smallest and m largest forecasts and sets them as the $(m+1)$th smallest and $(m+1)$th largest forecasts.
- Median-based combination: takes the median of forecast results from all models so that the prediction is more robust to extreme values.
- Error-based combining: sets weight ω_k to be the inverse of the past forecast error of the model k, for $k = 1, 2, \ldots, K$.

12.2   R Package "forecastHybrid"

12.2.1   Installation

In the R environment, hybrid models can also be identified using the hybridModel() function included in the package forecastHybrid[1]. This package

---

[1] https://cran.r-project.org/web/packages/forecastHybrid/forecastHybrid.pdf

provides convenient functions for ensemble forecasts in R combining approaches from the "forecast" package. The "forecastHybrid" package provides user-friendly functions to ensemble forecast results from available approaches in the "forecast" package.

The stable release of the "forecastHybrid" package is hosted on CRAN and can be installed as usual.

```
install.packages("forecastHybrid")
# Load the package
library(forecastHybrid)
```

As aforementioned, we can use the "devtools" package to install the latest developed version.

```
devtools::install_github("ellisp/forecastHybrid/pkg")
```

12.2.2 An Introduction

For the approaches in the "forecast" package, the following models are incorporated:

- `auto.arima()`: the best ARIMA model selected by either AIC, AICc or BIC criterion;
- `ets()`: exponential smoothing state space model;
- `stlm()`: applies an STL decomposition and models the seasonally adjusted data using the model passed as the model function or specified using method;
- `nnetar()`: feed-forward neural networks with a single hidden layer and lagged inputs;
- `tbats()`: TBATS model (exponential smoothing state space model with Box-Cox transformation, ARMA errors, trend and seasonal components) proposed by De Livera et al. (2011);
- `thetaf()`: the theta method proposed by Assimakopoulos and Nikolopoulos (2000), which is equivalent to simple exponential smoothing with drift;
- `snaive()`: the seasonal naive method (or the random walk method).

These methods can be combined using combination methods introduced in Section 12.1, e.g., equal weights, weights based on in-sample errors (Bates and Granger, 1969), or weights from cross-validation. To evaluate the model accuracy, cross-validation is also implemented with user-supplied forecasting models.

We consider the daily new death count for Florida. We illustrate the usage of
opera using the daily new deaths time series for the state of Florida. First, we
load the data.

```
library(tidyverse)
# devtools::install_github('FIRST-Data-Lab/IDDA')
library(IDDA)
data(state.ts)

Florida.ts <- state.ts %>%
  dplyr::filter(State == "Florida")
head(Florida.ts)
## # A tsibble: 6 x 9 [1D]
## # Key:       State [1]
## # Groups:    State [1]
##     State    Region Division        pop DATE       Infected
##     <chr>    <fct>  <fct>         <int> <date>          <dbl>
## 1 Florida South  South Atla~ 2.13e7 2020-01-23           0
## 2 Florida South  South Atla~ 2.13e7 2020-01-24           0
## 3 Florida South  South Atla~ 2.13e7 2020-01-25           0
## 4 Florida South  South Atla~ 2.13e7 2020-01-26           0
## 5 Florida South  South Atla~ 2.13e7 2020-01-27           0
## 6 Florida South  South Atla~ 2.13e7 2020-01-28           0
## # ... with 3 more variables: Death <int>,
## #   Y.Infected <dbl>, Y.Death <int>
```

When using the "forecastHybrid" package, it is better to store the time series
as a "ts" object in R. Let us consider the time series of daily new death counts
for Florida, and we first turn this into a "ts" object using the ts() function.
Since we have observed the 7-day cycle for this data, we can simply add a
frequency = 7 argument. Read Seasonal periods[2] for an explanation of how
to set the frequency for a seasonal period in "ts" objects. We use all the days
in 2020 except the last 14 days as the training set and the last 14 days as the
test set.

```
n <- length(Florida.ts$Y.Death)
h <- 14
y <- ts(Florida.ts$Y.Death[1:(n-h)], start = c(2020, 1), frequency = 7)
hmod <- hybridModel(y)
hmod.fc <- forecast(hmod, h = h)
```

---

[2]https://robjhyndman.com/hyndsight/seasonal-periods/

```r
# Use autoplot to display the time series and the forecasting results.
# autoplot(forecast(hmod), xlab = "Days", ylab = "Daily new deaths",
#     main = "Combined forecast from different models")
hmod.fc.report <- as.data.frame(hmod.fc) %>%
  mutate(DATE = as.Date("2020-12-17") + 1:14)
names(hmod.fc.report) <- c("Y.Death", "Lo.80", "Hi.80",
                      "Lo.95", "Hi.95", "DATE")

ggplot(Florida.ts, aes(DATE, Y.Death)) +
  geom_line(aes(color = 'Reported')) +
  labs(x = "Days", y = "Daily new deaths") +
  # Add prediction intervals
  geom_ribbon(mapping = aes(x = DATE,
                            y = Y.Death,
                            ymin = Lo.95,
                            ymax = Hi.95,
                            fill = '95% Prediction intervals'),
              data = hmod.fc.report, alpha = 0.4) +
  # Add line for predicted values
  geom_line(mapping = aes(x = DATE,
                          y = Y.Death,
                          color = 'Predicted'),
            data = hmod.fc.report,
            key_glyph = "timeseries") +
  scale_color_manual(values = c(Reported = "black",
                                Predicted = "red")) +
  scale_fill_manual("", values = "pink") +
  guides(color = guide_legend(title = "Series"),
         fill = guide_legend(title = "")) +
  theme(legend.position = "bottom")
```

The "hybridModel" object stores the individual component models, which can be viewed in corresponding plots. As introduced before, all the forecast models in the "forecast" package are applicable here. For example, the following example displays how to visualize the forecast from the auto.arima model.

```r
# Examine the individual models
hmod$auto.arima
```

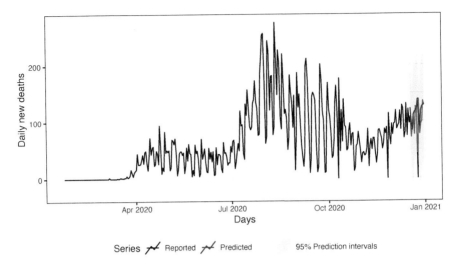

FIGURE 12.2: Combined forecast from different models.

```
## Series: y
## ARIMA(1,0,1)(0,1,1)[7]
##
## Coefficients:
##           ar1      ma1     sma1
##         0.961   -0.793   -0.570
## s.e.    0.020    0.040    0.059
##
## sigma^2 estimated as 864:  log likelihood=-1550
## AIC=3108    AICc=3108    BIC=3123
```

```
arima.fc <- forecast(hmod$auto.arima)
```

```
arima.fc.report <- as.data.frame(arima.fc) %>%
  mutate(DATE = as.Date("2020-12-17") + 1:14)
names(arima.fc.report) <- c("Y.Death", "Lo.80", "Hi.80",
                            "Lo.95", "Hi.95", "DATE")
ggplot(Florida.ts, aes(DATE, Y.Death)) +
  geom_line(aes(color = 'Reported')) +
  labs(x = "Days", y = "Death Count") +
  # Add prediction intervals
```

```r
geom_ribbon(mapping = aes(x = DATE,
                          y = Y.Death,
                          ymin = Lo.95,
                          ymax = Hi.95,
                          fill = '95% Prediction intervals'),
            data = arima.fc.report, alpha = 0.4) +
# Add line for predicted values
geom_line(mapping = aes(x = DATE,
                        y = Y.Death,
                        color = 'Predicted'),
          data = arima.fc.report,
          key_glyph = "timeseries") +
scale_color_manual(values = c(Reported = "black",
                              Predicted = "red")) +
scale_fill_manual(values = "pink")   +
guides(color = guide_legend(title = "Series"),
       fill =  guide_legend(title = "")) +
theme(legend.position = "bottom")
```

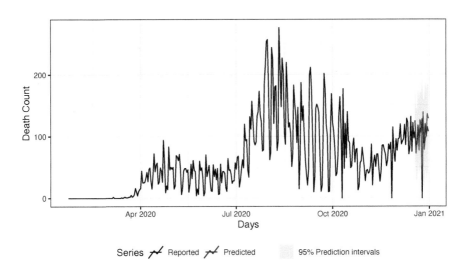

FIGURE 12.3: Forecast from the auto.arima model.

12.2.3   Model Diagnostics

We can use `print()` and `summary()` to export the information stored in the object generated from the `hybridModel()` function, which is an S3 object of the class "forecastHybrid."

```
print(hmod)
## Hybrid forecast model comprised of the following models:
 ↪  auto.arima, ets, thetam, nnetar, stlm, tbats
## ############
## auto.arima with weight 0.167
## ############
## ets with weight 0.167
## ############
## thetam with weight 0.167
## ############
## nnetar with weight 0.167
## ############
## stlm with weight 0.167
## ############
## tbats with weight 0.167
summary(hmod)
##              Length Class          Mode
## auto.arima  18      forecast_ARIMA list
## ets         19      ets            list
## thetam      23      thetam         list
## nnetar      15      nnetar         list
## stlm         9      stlm           list
## tbats       25      tbats          list
## weights      6      -none-         numeric
## frequency    1      -none-         numeric
## x          330      ts             numeric
## xreg         3      -none-         list
## models       6      -none-         character
## fitted     330      -none-         numeric
## residuals  330      ts             numeric
```

Figure 12.4 displays a plot with the actual and fitted values of each component model on the data.

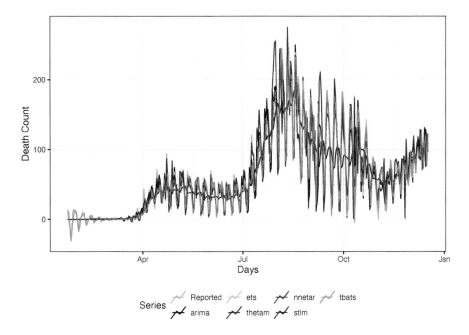

FIGURE 12.4: Reported counts and fitted values from each component model.

12.2.4 Forecasting

Forecasting is of great importance in time series modeling. We use the `fore-cast()` function to produce a forecast for the next 14 days using the ensembling method. The prediction intervals are preserved from the individual component models. Since the most extreme values from an individual model are used to form the intervals, the ensemble prediction intervals could be somewhat conservative.

```
hfc <- forecast(hmod, h = 14)
# View the point forecasts
hfc$mean
## Time Series:
## Start = c(2067, 2)
## End = c(2069, 1)
## Frequency = 7
## [1] 123.12  96.80  84.33 111.51 116.33 118.19 121.26
## [8] 139.20 102.34  90.28 119.23 126.10 128.34 128.92
# View the upper prediction interval
```

```
hfc$upper
##            80%    95%
##   [1,] 191.7 215.1
##   [2,] 147.3 171.1
##   [3,] 127.3 147.3
##   [4,] 156.7 177.5
##   [5,] 183.6 208.6
##   [6,] 170.3 195.7
##   [7,] 186.1 211.8
##   [8,] 198.6 224.7
##   [9,] 153.6 180.0
## [10,] 148.5 160.1
## [11,] 189.9 192.9
## [12,] 205.8 217.6
## [13,] 215.7 218.3
## [14,] 209.2 220.6
# View the lower prediction interval
hfc$lower
##            80%     95%
##   [1,] 75.316  55.725
##   [2,] 43.767  23.199
##   [3,]  5.667 -18.538
##   [4,] 23.592  -1.010
##   [5,] 58.528  37.596
##   [6,] 52.130  30.824
##   [7,] 65.386  44.077
##   [8,] 67.751  44.155
##   [9,] 36.504  12.091
## [10,]  1.239 -25.619
## [11,] 19.447  -7.769
## [12,] 52.590  27.339
## [13,] 45.080  20.042
## [14,] 58.975  33.464
```

Now, we plot the forecast for the next 14 days with the prediction intervals; see Figure 12.5.

```
hfc.report <- data.frame(Lo.95 = hfc$lower[,2],
                         Hi.95 = hfc$upper[,2], Y.Death = hfc$mean) %>%
  mutate(DATE = as.Date("2020-12-17") + 1:14)
ggplot(Florida.ts, aes(DATE, Y.Death)) +
  geom_line(aes(color = 'Reported')) +
```

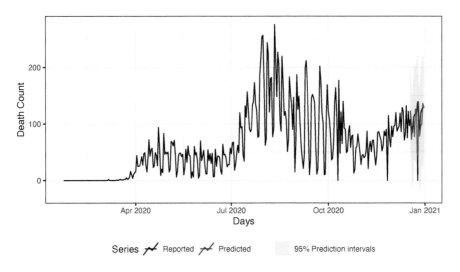

FIGURE 12.5: Two-week-ahead ensemble forecast from different methods.

```
labs(x = "Days", y = "Death Count") +
# Add prediction intervals
geom_ribbon(mapping = aes(x = DATE,
                          y = Y.Death,
                          ymin = Lo.95,
                          ymax = Hi.95,
                          fill = '95% Prediction intervals'),
            data = hfc.report, alpha = 0.4) +
# Add line for predicted values
geom_line(mapping = aes(x = DATE,
                        y = Y.Death,
                        color = 'Predicted'),
          data = hfc.report,
          key_glyph = "timeseries") +
scale_color_manual(values = c(Reported = "black",
                              Predicted = "red")) +
scale_fill_manual(values = "pink") +
guides(color = guide_legend(title = "Series"),
       fill = guide_legend(title = "")) +
theme(legend.position = "bottom")
```

After the model fitting, one can check the weights that are stored in the `weights` attribute. The weight can be replaced by the user, with retaining the component name. Note that to guarantee the unbiasedness of the forecasts, the vector of the weights should add up to unity. To ensure the forecasts remain unbiased, we suggest the user use a similar weights scale. In addition, since weights are not assigned by position but by component name, the vector that replaces weights must retain names specifying the component model it corresponds to.

```
hmod$weights
## auto.arima        ets      thetam      nnetar         stlm
##     0.1667     0.1667      0.1667      0.1667       0.1667
##      tbats
##     0.1667
```

12.2.5   Performing Cross-Validation on a Time Series

To evaluate a forecasting model's out-of-sample performance, cross-validation is a useful tool and has been widely adopted in independent datasets. However, compared to those regular leave-one-out or k-fold cross-validation for independent data tools, the serial correlation in time series makes it more complicated to perform cross-validation.

Here, we use the cvts() function to perform cross-validation on time series, which tackles the non-dependence issue on observations by utilizing two methods: (i) a non-rolling cross-validation approach, where a fixed training window is moved forward with the forecast horizon for each iteration; (ii) a rolling approach, where a rolling training window is adopted with growing by one observation each round until the entire series is captured by the training window and the forecast horizon.

We illustrate the usage of the cvts() function by the following example, where we compare a stlm model and a naive model, the latter treated as a good baseline. The arguments that need attention include:

- rolling: if TRUE, non-overlapping windows of size maxHorizon will be used for fitting each model; otherwise, the size of the dataset used for training will grow by one each iteration during cross-validation;
- windowSize: the starting length of the time series to fit a model;
- maxHorizon: the forecast horizon for predictions from each model.

Since a naive forecast is a good baseline that any decent model should surpass, let's see how the stlm model compares with the naive model.

```
stlmMod <- cvts(y, FUN = stlm, windowSize = 100, maxHorizon = 7)
naiveMod <- cvts(y, FUN = naive, windowSize = 100, maxHorizon = 7)
```

```
accuracy(stlmMod)
##                           ME   RMSE    MAE
## Forecast Horizon  1  0.8081  32.96  25.55
## Forecast Horizon  2 -1.0981  42.80  24.46
## Forecast Horizon  3 -0.5347  36.99  27.34
## Forecast Horizon  4 -2.9073  38.19  27.54
## Forecast Horizon  5 -1.2949  31.73  24.28
## Forecast Horizon  6 -5.1584  51.69  35.24
## Forecast Horizon  7 -1.6227  40.29  29.70
accuracy(naiveMod)
##                           ME   RMSE    MAE
## Forecast Horizon  1 -20.781  39.27  28.47
## Forecast Horizon  2 -58.062  74.62  61.88
## Forecast Horizon  3 -52.719  68.61  54.09
## Forecast Horizon  4   5.062  35.97  27.19
## Forecast Horizon  5   5.906  35.24  27.53
## Forecast Horizon  6   6.562  52.12  37.19
## Forecast Horizon  7   2.375  39.80  27.31
```

Note that the above errors are based on each of the forecast horizons. Suppose we are interested in reporting the average errors over all forecast horizons up to maxHorizon instead of producing metrics for each individual horizon. We can specify horizonAverage=TRUE.

```
stlmMod.mean <- cvts(y, FUN = stlm, windowSize = 100,
                     maxHorizon = 7, horizonAverage = TRUE)
naiveMod.mean <- cvts(y, FUN = naive, windowSize = 100,
                      maxHorizon = 7, horizonAverage = TRUE)
accuracy(stlmMod.mean)
##                         ME   RMSE    MAE
## Forecast Horizon  1 -1.687  23.43  15.62
accuracy(naiveMod.mean)
##                         ME   RMSE    MAE
## Forecast Horizon  1 -15.95  31.64  21.82
```

12.2.6 Weights Selection Using Cross-Validation

In previous sections, we explored fitting hybridModel() objects with weights = "equal". Cross-validation can also be applied here to select appropriate weights. The package allows the user to leverage the process conducted in cvts() to select the appropriate weights intelligently. The method is based on the expected out-of-sample forecast accuracy of each component model. While this weight selection procedure is methodologically sound, it also comes at a significant computational cost. In addition to the final fit on the entire dataset, the cross-validation procedure involves fitting each model many times for each cross-validation fold. To address this issue, parallel computing is adopted here if multiple cores are available in the user's computer system. Some of the arguments explained above in `cvts()`, such as `windowSize` and the `cvHorizon`, can also be controlled here.

```
cvMod <- hybridModel(y, weights = "cv.errors", windowSize = 100,
                     cvHorizon = 8, num.cores = 4)
cvMod
```

12.3 R Package "opera"

The R Package "opera" (online prediction by expert aggregation) provides methods to perform robust online predictions, for regression-oriented time series, by combining a set of user-provided forecasts.

12.3.1 Installation

We can install "opera" with:

```
install.packages("opera")
# Load the package
library(opera)
```

We can also install the development version of opera with the package "devtools":

```
install.packages("devtools")
devtools::install_github("dralliag/opera")
```

12.3.2 An Introduction

There are three important functions provided by "opera":

- `mixture()`: builds the algorithm object;
- `predict()`: makes a prediction by using the algorithm;
- `oracle()`: evaluates the performance of the experts and compares the performance of the combining algorithm.

We illustrate the usage of opera using the daily new deaths time series for the state of Florida. First, we load the data, and we divide it into a training set and a test set. The training data will be used to build the base forecasting models. The test set will be used to evaluate the performance and to run the combining algorithms.

```
library(IDDA)
data(state.ts)

# Florida daily new deaths
Florida.ts <- state.ts %>%
  dplyr::filter(State == "Florida") %>%
  dplyr::select(Infected, Death, Y.Infected, Y.Death)
head(Florida.ts)
```

```
## # A tsibble: 6 x 6 [1D]
## # Key:        State [1]
## # Groups:     State [1]
##    State    Infected Death Y.Infected Y.Death DATE
##    <chr>       <dbl> <int>      <dbl>   <int> <date>
## 1 Florida         0     0          0       0 2020-01-23
## 2 Florida         0     0          0       0 2020-01-24
## 3 Florida         0     0          0       0 2020-01-25
## 4 Florida         0     0          0       0 2020-01-26
## 5 Florida         0     0          0       0 2020-01-27
## 6 Florida         0     0          0       0 2020-01-28
```

```
library(tsibble)
# Set training data from JAN 23 to DEC 17, 2020
train <- Florida.ts %>%
  filter_index("2020-01-23" ~ "2020-12-17")
# Set test data from DEC 18 to DEC 31, 2020
test <- Florida.ts %>%
  filter_index("2020-12-18" ~ "2020-12-31")
h <- nrow(test)
```

We consider two base time series models, ARIMA and ETS, to be combined later.

```
# ETS modeling and forecasting
ETS_fit <-  train %>%
  model(ETS(Y.Death ~ error("A") + trend("A") + season("A")))
ETS <- forecast(ETS_fit, h = h)
# ARIMA modeling and forecasting
ARIMA_fit <-  train %>%
  model(ARIMA(Y.Death, stepwise = FALSE))
ARIMA <- forecast(ARIMA_fit, h = h)
```

Next, we will build the expert matrix and the time series to be predicted. Then, we aggregate the experts using one of the possible aggregation procedures. To start the procedure, we need to initialize the algorithm by defining the type of algorithm (e.g., ridge regression, exponentially weighted average forecaster, fixed-share aggregation, polynomial Potential aggregation), the possible parameters, and the evaluation criterion (square, absolute, percentage, or pinball). Below, we use the ML-Poly algorithm, evaluated by the square loss.

```
library(opera)
death.fc <- cbind(ETS = ETS$.mean, ARIMA = ARIMA$.mean)
MLpol0 <- mixture(model = "MLpol", loss.type = "square")
```

We can also check the weights assigned to each individual component model when combining the forecasts.

```
library(opera)
weights <- predict(MLpol0, death.fc, test$Y.Death, type = 'weights')
head(weights)
##            ETS  ARIMA
## [1,] 0.50000 0.5000
## [2,] 0.00000 1.0000
## [3,] 0.00000 1.0000
## [4,] 0.00000 1.0000
## [5,] 0.05613 0.9439
## [6,] 0.00000 1.0000
```

Finally, we perform the predictions by using the predict function, and make the plot.

```
death.cfc <- predict(MLpol0, death.fc, test$Y.Death, type =
⤷ 'response')
# Make the plot
death.cfc.report <- as.data.frame(death.cfc) %>%
  mutate(DATE = as.Date("2020-12-17") + 1:14)
names(death.cfc.report) <- c("Y.Death", "DATE")
ggplot(Florida.ts, aes(DATE, Y.Death)) +
  geom_line(aes(color = 'Reported')) +
  labs(x = "Days", y = "Number of deaths") +
  geom_line(mapping = aes(x = DATE,
                          y = Y.Death,
                          color = 'Predicted'),
            data = death.cfc.report,
            key_glyph = "timeseries") +
  scale_color_manual(values = c(Reported = "black",
                                Predicted = "red")) +
  guides(color = guide_legend(title = "Series"),
         fill = guide_legend(title = "")) +
  theme(legend.position = "bottom")
```

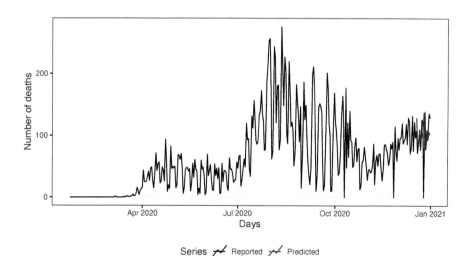

FIGURE 12.6: Two-week-ahead ensemble forecast from ETS and ARIMA models using "opera."

12.4 Further Reading

For further information on the R packages used in this chapter, readers can consult the following web resources.

- `https://cran.r-project.org/web/packages/forecastHybrid;`
- `https://cran.r-project.org/web/packages/forecastHybrid/vignettes/fo recastHybrid.html;`
- `https://cran.r-project.org/web/packages/opera.`

12.5 Exercises

Predicting the severity and speed of disease transmission is crucial to resource management and developing strategies. This project focuses on the projection of COVID-19. You will work with the `state.long` dataset in the IDDA package, which contains some COVID-19 count information from January 22 to December 31, 2020.

```
library(IDDA)
data(state.long)
```

Note that `state.long` is a data frame with 16,905 rows and seven columns. Each row of the data frame stands for one state on a specific date. The columns are `State`, `Region`, `Division`, `pop`, `DATE`, `Infected` and `Death`. See Appendix B for a detailed explanation of these variables.

You will make a projection of the number of daily new death cases and the cumulative number of death cases for each state in the US. For each state, you will hold out the last two weeks as test data and use all the data except the last two weeks as your training data. Based on the training data, you fit the model and make the projection of the number of daily new death cases and the cumulative number of death cases for the last two weeks.

Recall that we have learned three main types of epidemic models: (i) mechanistic models, (ii) phenomenological models, and (iii) hybrid models. You can choose some models (with a minimum of five models and a maximum of ten models) from what we have discussed in class or those in the "forecastHybrid" R package as individual models to make the forecast, and combine these models to make an ensemble forecast.

For each state and forecast method, you need to provide a point forecast, 80% prediction intervals, and 90% prediction intervals for the last two weeks. Then you aggregate all the state forecasts to obtain the forecast for the US.

To demonstrate your projection results, you need to develop a shiny app. Through the app, you can (i) select each state and showcase the corresponding time series plot, (ii) select the forecast method, and illustrate your point projection, 80% prediction intervals and 90% prediction intervals.

A

Appendix A

This appendix introduces the basics of the R language and the RStudio programming environment. We will also go over some primary programming skills in R, such as downloading, reading, manipulating, importing/exporting data, and some basic control structures in R.

A.1 R Introduction and Preliminaries

A.1.1 The R Environment and Language

R is an integrated suite of software facilities for data manipulation, calculation, and graphical display. It is the most popular language in the world of Data Science.

Why Choose R for Data Science

- R is an open-source programming language, and a free software under the terms of the Free Software Foundation's GNU General Public License in source code form. R is maintained by a community of active users.

- R has powerful visualization libraries such as "ggplot2" and "plotly," and it can produce stunning graphs and visualizations.

- R's language has a powerful, easy-to-learn syntax with many built-in statistical functions.

- R provides excellent capabilities to build aesthetic web applications.

- R provides various important packages for data wrangling like "dplyr," "purrr," "readxl," "googlesheets," "tidyr," etc., which makes data wrangling a lot easier.

- With R, users can apply statistical models and machine learning algorithms to gain insights into future events.

A.1.2 Obtaining R, RStudio and Installation

We can obtain sources, binaries, and documentation for R via the Comprehensive R Archive Network (CRAN). The current CRAN members are listed at `http://cran.r-project.org/mirrors.html`, and we can choose the preferred CRAN mirror to download R. Then, we need to select the operating system, and installation files are available for the Windows, Mac, and Linux operating systems. If downloading R for Windows, we will be asked to select from the "base" or "contrib" distributions, and we will select the base distribution. Installation is straightforward, and we just need to follow the instructions.

As an integrated development environment for R, RStudio is available in open source and commercial editions and runs on the desktop (Microsoft Windows, Linux, Apple MacOS). After installing R, we then install RStudio, which can be downloaded and installed from RStudio at `https://www.rstudio.com/products/rstudio/download/`. Click on the download link corresponding to the computer's operating system.

A.2 Starting RStudio

RStudio is most easily used in an interactive manner. After installing R and RStudio, there will be two new programs (also called applications) that we can open.

Let's open RStudio. After opening it, we will see the RStudio interface to R as seen in Figure A.1 below, with four panes dividing the screen: the source pane, the console pane, the files pane, and the environment pane.

- **Source:** This is where we write the code that needs to be run. It contains all of the individual R scripts.

- **Console:** This is the place where R actually evaluates code.

- **Environment and History:** This shows the list of the variables and objects in R that are currently available in the working environment. There are also other tabs within the environment window, such as "history," which contains a history of all code typed in the past. It also has a tab called "connection." It shows all the connections that have been made to supported data sources.

- **Viewer:** Unlike the other three panes, this pane has several tabs nested within it. The "files" tab shows all the files and folders in the current directory, which the program points to right next to the home icon below the header for the pane. The "plots" tab displays and allows for saving any graphics that we have created. The "packages" tab exhibits all the packages

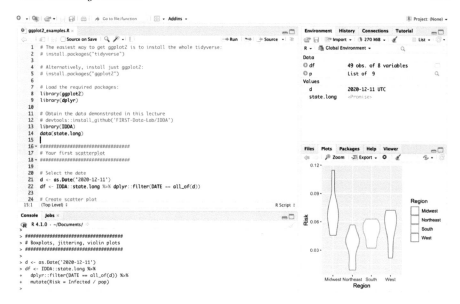

FIGURE A.1: RStudio interface to R.

that are currently installed. The "help" tab allows searching the R documentation for help.

A.2.1 Source Pane

The source pane is where we create and edit the R Scripts. The source editor in RStudio includes a variety of productivity-enhancing features, such as syntax highlighting, code completion, multiple-file editing, and find/replace.

It can also help us open, edit and execute these programs. For example, to open an existing file, we use either the **File -> Open File...** menu or the **Recent Files** menu to select from recently opened files. To create a new script, we click on the **New Document** button in the top left and select "R Script."

We can execute R code directly from the source pane by the following (for Windows machines; for Mac machines, replace **Ctrl** with **Cmd**).

- To run a single line, move the cursor anywhere in that line and press the key **Ctrl+Enter**.

- To run more than one line at a time, highlight all the code to be executed, and then press the key **Ctrl+Enter** or click the **Run** button.

- To run the entire script, press the **Ctrl+Shift+Enter** key.

To save the code as a program outside of R, select **File -> Save** and make
sure to use an `.R` extension on the file name.

A.2.2   Console Pane

To have R actually evaluate the code, we need to first deliver the code to
the console pane. At the beginning of the console, we see a symbol >, which
is called the **prompt**. In what follows below, it is not typed but is used to
indicate where to type.

The console pane can be used as a calculator. Below are some examples:

```
2 + 15
## [1] 17
(12 - 3) / 6
## [1] 1.5
exp(2 ^ 2)
## [1] 54.6
cos(pi / 2)
## [1] 6.123e-17
log(1)
## [1] 0
```

Results from these kinds of calculations can be stored in an object, the entity
that R operates on; see Venables et al. (2009). Let's consider the following R
code, where the <- is used to make the assignment and is read as "gets."

```
x <- exp(2 ^ 2)
x
```

```
## [1] 54.6
```

```
alpha <- log(1:5)
alpha
```

```
## [1] 0.0000 0.6931 1.0986 1.3863 1.6094
```

```
temp <- alpha * x
temp
```

```
## [1]  0.00 37.84 59.98 75.69 87.87
```

```
county <- c("LosAngeles", "Lake", "SanFrancisco")
county
```

```
## [1] "LosAngeles"    "Lake"          "SanFrancisco"
```

The function `c()` can take an arbitrary number of vector arguments and whose value is a vector that is created by combining its arguments.

To see a listing of the objects, we can call either the `ls()` or the `objects()` function. To delete an object, we can use `rm()` and insert the object name in the parentheses, for example, `rm(temp, county)`. The `rm(list=ls())` can be used to clean out all objects from the workspace.

One can permanently store any objects created during an R session in a file for future R sessions. To save all the currently available objects, we can write the objects to a file with the ".RData" format in the current directory (**Save Workspace**), and the command lines used in the session are saved to a file with the ".Rhistory" format (**Save History**). The following code

```
save(x, alpha, file = "objects.RData")
```

saves objects x and `alpha` to the file "objects.RData" in the current working directory. We can use `load("objects.RData")` to load the objects in file "objects.RData."

A.2.3 Error Messages

R can provide intuitive error messages regarding the submitted syntax, and these comments are printed in the console. For example, if an object or function is spelled incorrectly, we will see an error like "Error: object 'X' not found" or "Error: could not find function." R is case-sensitive, so we will receive an error if an uppercase letter X is used, but we should have typed a lowercase letter.

```
# An uppercase letter X is used,
# but we should have typed a lowercase letter
X + 2
```

Another common error is that a closing parenthesis, bracket, or quotation is missing. We will also receive an error if we forget to add an operator or pipe (%>%) after the phrase before moving on to the next indented line of continued code.

```
# Forget a closing parentheses or bracket
# Forget the operator between 3 and z
2 + (3xz

# Forget to add a pipe (%>%) after the phrase before
state.ts <- state.long %>%
  group_by(State)
  mutate(YDA_Infected = lag(Infected, order_by = DATE))
```

A.2.4   R Help

To see a listing of all built-in R functions, open the Help by selecting **Help -> R Help** from the main menu bar. Under **Reference**, select the link called **Packages**. All built-in R functions are stored in a package. By clicking "Packages," we can scroll down to find help on the median() function in the R "stats" package. Note the full syntax for median() is

```
median(x, na.rm = FALSE, ...)
```

Two arguments used in this function are:

- x: an object for which a method has been defined, or a numeric vector containing the values whose median is to be computed;

- na.rm: a logical value indicating whether NA values should be stripped before the computation proceeds.

If we know the exact name of the function, we can simply type help(function name) at the R console command prompt to bring up its help in a window

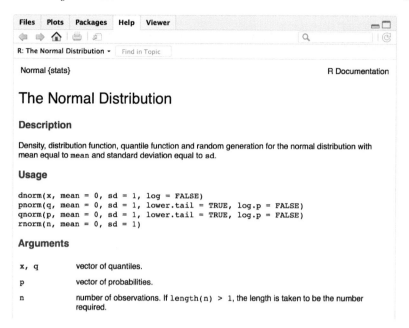

FIGURE A.2: The R help for the R function `pnorm()`.

inside of R. For example, `help(pnorm)` brings up Figure A.2. An alternative is `?pnorm`.

For a non-standard name (such as the Pipe operator in package "magrittr"), the argument must be enclosed in double or single quotes, making it a "character string" like `help('%>%')` or `?"%>%"`. Sometimes, we may not know the exact name of a function, or we are not sure of its existence. There is a very useful function called `apropos('argument')` that lists all functions that contain the "argument" as part of their names. Note that the argument must be put within either single or double quotation marks. For example, here is what we get when looking for similar functions containing the string `predict`:

```
apropos('predict')
```

```
##   [1] "makepredictcall"
##   [2] "napredict"
##   [3] "predict"
##   [4] "predict.bam"
##   [5] "predict.gam"
##   [6] "predict.glm"
##   [7] "predict.lm"
```

```
##  [8] "Predict.matrix"
##  [9] "Predict.matrix.Bspline.smooth"
## [10] "Predict.matrix.cr.smooth"
## [11] "Predict.matrix.cs.smooth"
## [12] "Predict.matrix.cyclic.smooth"
## [13] "Predict.matrix.duchon.spline"
## [14] "Predict.matrix.gp.smooth"
## [15] "Predict.matrix.mrf.smooth"
## [16] "Predict.matrix.pspline.smooth"
## [17] "Predict.matrix.random.effect"
## [18] "Predict.matrix.sf"
## [19] "Predict.matrix.soap.film"
## [20] "Predict.matrix.sos.smooth"
## [21] "Predict.matrix.sw"
## [22] "Predict.matrix.t2.smooth"
## [23] "Predict.matrix.tensor.smooth"
## [24] "Predict.matrix.tprs.smooth"
## [25] "Predict.matrix.ts.smooth"
## [26] "Predict.matrix2"
## [27] "PredictMat"
```

Note that the argument is a string, so it does not need to be an actual word or name of a function. For example, `apropos('pred')` will return the same results.

Sometimes we want to learn about all functions involving a certain term, but searching for R-related pages on that term returns too many irrelevant results. This term may not even be an R function or command, making the Google search all the more difficult, even with good searching techniques. In these situations, use the `help.search('argument')` function. Again, we need to put the arguments within single or double quotation marks. This will return all functions with the "argument" in the help page title or as an alias.

A.2.5   R Packages

In SAS/SPSS installations, we usually have everything we have paid for installed at once. R provides a modular platform, and its main installation will install R and a popular set of **libraries**, a collection of R functions, compiled code and sample data. Hundreds of other libraries are available to install separately from the CRAN.

The most common functions used in R are contained within the "base" package. If we want to use functions in other packages, we need to install and then load the package into R. Packages, if they have already been downloaded from a CRAN mirror site, can be loaded using this procedure. If the package has not been downloaded, it can be installed using the `install.packages(package name)` option. Also, an installed package can be loaded by specifying `library(name of the package)`.

For example, to install the "ggplot2" package for data visualization, we can select **Tools -> Install Packages** from the main menu. A number of locations around the world will come up. Insert the "ggplot2" package and select **Install**. The package will now be installed. This only needs to be done once on the computer. To load the package into the current R session, we can call the `library(ggplot2)` at the R Console prompt. Note that this only needs to be done once in an R session. If R was closed, we would need to use the `library()` function again when reopening R.

We can also use the viewer pane to install and update packages. In the viewer pane, the tab for **packages** shows all the installed and available packages. To load the package for use, we can click on the checkbox next to the name of the package, or use the command `library(package name)`, which will appear in the console pane of the interface. Clicking on the package's name brings up the description of what the package does. If the package contains example datasets, we can load them with the `data` command. We can enter `data()` to see what is available and then `data(mydata)` to load one named, for example, `mydata`.

The "Packages" tab also contains a link to the package name, which provides an overview of what the package does. For example, next to the package "ggplot2," we see the package title, which states, "Create Elegant Data Visualisations Using the Grammar of Graphics." Thus, "ggplot2" is a package that makes elegant data visualizations. Clicking on the package link provides an alphabetized list of all the package does, such as a function called `aes()` that helps construct aesthetic mappings. Another function called `borders()` helps create a layer of map borders. The link to `aes()` explains what the function is about and the syntax it uses. Understanding the structure of this description helps us understand how to use packages.

A.2.6 Creating a Project and Setting a Working Directory

Before creating a dataset, we need an idea of where the files are stored with R and how to manipulate those files. Each R session has a default location where R will look for files to load, called the **working directory**. If we don't set it to the desired location, we could easily write files to an undesirable file location.

We can only have one working directory active at any given time. A simple way to check on the current working directory is to type the command `getwd()` into the command console. It tells us what the current working directory is:

```
getwd()
```

It is very important to ensure that the working directory is where we want the created files and objects to be stored. If it is not the intended directory to use, the simplest way to change it is by using the "viewer" pane and clicking on the tab "More." To change the working directory, we can also use the `setwd()` function with the working directory entered as a character string. This example shows how to change the working directory to a folder called "D:/EpiStat/Project1":

```
setwd("D:/EpiStat/Project1")
getwd()
```

Note that the separator between folders is /, as it is on Linux and Mac systems. When working in Windows, we need to either use / or \\.

An RStudio project is mainly a folder that will hold everything associated with a specific project, such as data, history of commands used, objects, or variables that are created. To create a new project by opening RStudio, we can click on the **File -> New Project**. As shown in Figure A.3, a new window opens and asks if we would like to open the project in an existing directory, a new one, or simply version control. We may either choose an existing directory or a new one. However, if a new one is chosen, we need to ensure it is selected as the working directory, as shown earlier.

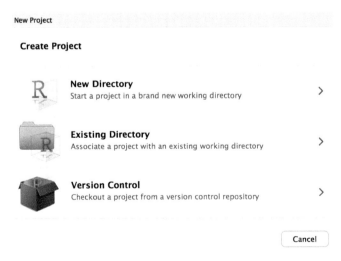

FIGURE A.3: Interface for creating a project in R.

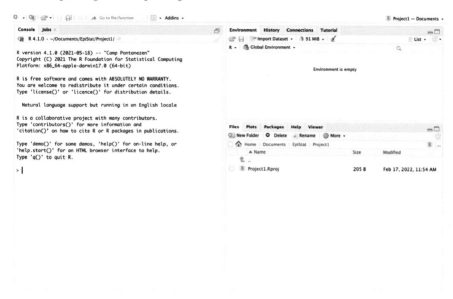

FIGURE A.4: Directory with new project.

Once the project is created, we will see a ".proj" file under the files section in the viewer pane, as shown in Figure A.4. As the figure shows, a new project called "Project1.rproj" has been created in the directory "~Documents/EpiStat/Project1." By setting it as the working directory, any work done will now be stored in this directory and in this project as long as it is saved.

A.3 Exporting and Importing Data

In this section, we will introduce how to read plain-text rectangular files into R, and how to export data from R to .txt (tab-separated values) and .csv (comma-separated values), and .RData (R data) file formats.

A.3.1 Data Export

A.3.1.1 The `cat()` and `print()` Functions

The function `cat()` is useful for producing output in user-defined functions.

```
cat("Welcome to Data Science!","\n") #\n: newline
```

```
## Welcome to Data Science!
```

We can also print a character string in the file using the function `cat()` as follows.

```
cat(file = "test.txt", "123456", "987654", sep = "\n")
```

The function `print()` prints its argument. It is a generic function which means that new printing methods can be easily added for new classes.

```
print("Good morning!")
```

```
## [1] "Good morning!"
```

A.3.1.2   Writing a Data Frame to a File

One common task in R is to write a data frame to file, in which the data are presented in a rectangular grid, possibly with row and column labels. This can be handled by the functions `write.table()` and `write()`.

The following code writes the data of an R data frame object into the clipboard from where it can be pasted into other applications.

```
County <- c("Kent", "NewCastle", "Sussex")
Infection <- c(9173, 33211, 14936)
Death <- c(167, 462, 297)
DE.cases <- data.frame(County = County, Infection = Infection,
                       Death = Death)

# Write cases to a file named "DEcases.txt"
write.table(DE.cases, file = "DEcases.txt", sep = "\t",
            col.names = NA, quote = F)
```

The argument `col.names` guarantees that the titles align with columns when row names are exported. The argument `quote` can be a logical value (`TRUE` or

FALSE) or a numeric vector. If it is TRUE, any character or factor columns will be surrounded by double quotes. If it is FALSE, nothing is quoted.

We can also write a data frame to a tab-delimited text file.

```
# Install and load the package
if(!require('pgirmess')) install.packages('pgirmess')
library(pgirmess)
# Write DE.cases to a tab-delimited text file
pgirmess::write.delim(DE.cases, file = "DEcases.txt")
# The following provides same results as write.delim
write.table(DE.cases, file = "DEcases.txt", sep = "\t")
```

We can write the data to CSV files as follows:

```
write.csv(DE.cases, file = "DEcases.csv")
```

It is worth mentioning that write.table() is the multipurpose work-horse function in base R for exporting data. The functions write.csv() and write.delim() are special cases of write.table() in which the defaults have been adjusted for efficiency.

The best way to store objects from R is with .RData files, which are specific to R and can store more than one object within a single file.

To save selected objects into one .RData file, use the save() function.

```
# Save the object as a new .RData file
save(DE.cases, file = "data/DEcases.rda")
```

When we run save(a, b, c, file = "myobjects.RData") with specific objects a, b and c as arguments, all of those objects will be saved in a single file called myobjects.RData. For example, let's create a few objects corresponding to a medical study.

```
# Create two objects
patients.df <- data.frame(id = 1:5,
                          sex = c("m", "f", "m", "f", "m"),
                          # Systolic blood pressure
                          sbp = c(110, 80, 97, 102, 132))

sbp.by.sex <- aggregate(sbp ~ sex,
```

```
                        FUN = mean,
                        data = patients.df)
```

```
# Save two objects as a new .RData file
save(patients.df, sbp.by.sex, file = "data/patients.rda")
```

A.3.2   Data Import

A.3.2.1   The `load()` Function

To load an `.RData` file, that is, to import all of the objects contained in the `.RData` file into the current workspace, we can use the `load()` function. For example, to load the three specific objects that we saved earlier (`DE.cases` in `DEcases.rda`, and `patients.df`, `sbp.by.sex` in `patients.rda`), we can run the following:

```
# Load objects in rda files into my workspace
load(file = "data/DEcases.rda")
load(file = "data/patients.rda")
```

To load all of the objects in the workspace that we have saved to the data folder in a working directory named `projectnew.rda`, we can run the following:

```
# Load objects in projectnew.rda into my workspace
load(file = "data/projectnew.rda")
```

A.3.2.2   The `read.table()` Function

Most realistic data has more than a few rows, and it is not easily entered during an R session at the keyboard. Using R input facilities, we can simply read large data objects as values from external files, and the requirements are fairly strict and even rather inflexible.

The external file will typically have a special form to read an entire data frame directly: (i) the first line of the file should have a name for each variable in the data frame; (ii) each additional line of the file has its first item, a row label, and the values for each variable. By default, numeric items (except row labels) are read as numeric variables and non-numeric variables, such as name and gender in the above example, as factors.

The function `read.table()` can then be used to read in a file that contains tabular data into R. For example, the `patients.txt` file contains the data like the following:

```
name gender age sbp
Linda F 45 85
Jason M 65 95
Susan F 80 107
Mike  M 70 92
Judy  F 60 112
```

It can be imported using `df <- read.table("patients.txt", header = TRUE)`, where the `header=TRUE` option specifies that the first line is a line of headings, and hence no explicit row labels are given. We can specify more options like `colClasses` in the following, which provides a character vector of class names assigned to each column.

```
df <- read.table("patients.txt",
                  colClasses = c("character", "character",
                                 "integer", "integer"),
                  header = TRUE)
```

If the values are separated by commas or another "delimiter," we need to specify the delimiters. The file `infection.txt` contains the county-level weekly infected case count of a certain disease in a state.

```
County1,3,5,7,2,6
County2,0,2,4,5,3
County3,9,10,4,8,6
```

```
infection.df <- read.table("infection.txt", sep = ",")

# names() is to get or set the names of an object.
names(infection.df) = c("County", "Week1", "Week2", "Week3",
                  "Week4", "Week5")
print(infection.df)
```

The `sep` option specifies the character that is separating the fields, and `sep = ""` by default in `read.table`.

A.3.3 The `read.csv()` Function

Alternatively, we may have data from a comma-separated values (CSV) file. The simplest way to import data from a CSV file into R is through a file format. For example, we have a CSV file named data.csv. Then the R function read.csv() will import it into R as follows

```
x <- read.csv(file="data.csv")
```

CSV files can be comma-separated or tab-delimited or any other delimiter specified by parameter sep. If the parameter header is TRUE, then the first row will be treated as the names of the variables. R offers two built-in functions that we can use to quickly and easily read CSV files: read.csv and read.csv2.

```
read.csv(file, header = FALSE, sep = ",", quote = "\"", dec = ".",
         fill = TRUE, comment.char = "", ...)
read.csv2(file, header = TRUE, sep = ";", quote = "\"", dec = ",",
          fill = TRUE, comment.char = "", ...)
```

The difference between read.csv() and read.csv2() is the default field separator, as "," and ";", respectively. Below is a simple CSV file example.

```
     d1  d2  d3  d4  d5  d6  d7  d8
r1   1   0   1   0   0   1   0   2
r2   1   2   2   1   2   1   2   1
r3   0   0   0   2   1   1   0   1
r4   1   2   0   0   0   1   2   1
r5   0   2   1   1   1   0   0   0
r6   2   2   0   1   1   1   0   0
r7   2   2   0   1   1   1   0   1
r8   0   2   1   0   1   1   2   0
r9   1   0   1   2   0   1   0   1
r10  1   0   2   1   2   2   1   0
r11  1   0   0   0   1   2   1   2
r12  0   0   1   1   2   0   0   0
```

To read this file into the R workspace, we can simply use the following code.

```
x <- read.csv("readcsv.csv", header = T, dec = ".", sep = "\t")
is.data.frame(x)
```

A.3.4 The "readr" Package

Although the "base" R package provides several functions that can be used to import plain text files, we can also use the "readr" package to import data. Compared to the corresponding base functions, functions in the "readr" package offer a fast and friendly way to read rectangular data, such as CSV, tab-separated values (TSV), and fixed width formatted (FWF) data. We can install and load "readr":

```
# The easiest way to get readr is to install the whole tidyverse:
install.packages("tidyverse")

# Alternatively, install just readr:
install.packages("readr")

library(readr)
```

The "readr" package supports the following file formats with various `read_*` functions:

- `read_csv()`: comma-separated values files;
- `read_tsv()`: tab-separated values files;
- `read_delim()`: general delimited files;
- `read_fwf()`: fixed width files;
- `read_table()`: tabular files where columns are separated by white-space;
- `read_log()`: web log files.

We refer to Chapter 11 of Wickham and Grolemund (2016) for more details of how to use the "readr" package.

A.3.5 Importing an Excel File into R

We can import an Excel file using the "xlsx" package. There are two main functions for reading both "xls" and "xlsx" Excel files: `read.xlsx()` and `read.xlsx2()`, and the latter is faster on big files compared to the former function. To start, here is a template to use to import an Excel file into R:

```
library("xlsx")
read.xlsx(file, sheetIndex, header = TRUE)
read.xlsx2(file, sheetIndex, header = TRUE)
```

A.3.6   Accessing Built-in Datasets

There are around 100 datasets offered by the R package "datasets," and others are available in various R packages. To see the list of datasets currently available, we can type `data()`. All the datasets supplied with R are available directly by name.

```
AirPassengers
```

However, many packages still use the earlier convention in which data was also used to load a dataset into R, for example,

```
data(AirPassengers)
```

and this can still be used with the standard packages. To access data from a particular package, use the `package` argument, for example,

```
library(rpart)
data(package = "rpart")
# devtools::install_github('FIRST-Data-Lab/IDDA')
library(IDDA)
data(state.ts, package = "IDDA")
```

A.4   Control Structures in R

A.4.1   Grouped Expressions and Control Structures

R is an expression language, and its only command type is a function or expression which returns a result. R allows us to group several commands together in braces, for example,

```
{expr1 ; expr2 ;...; expr10}
```

and the value of the group is the result of the last expression, `expr10`, in the group evaluated.

Control structures in R allow us to control the execution flow of a series of R expressions. Using control structures, we can incorporate some "logic" into the R code, respond to the data's inputs or features, and execute different R expressions accordingly.

A.4.1.1 Conditional Execution: If Statements

The **if statement** tests a condition and acts on it. It is probably the most commonly used control structure in R. A typical **if-else statement** has the conditional construction of the following form:

```
if (condition) expr_1 else expr_2
```

where condition must evaluate to a single logical value, and the result of the entire expression is then evident.

Here is an example of an if-else statement:

```
if (x < 3) print("x is less than 3")
else print ("x is great than or equal to 3")
```

If statement A and statement B consist of more than one statement, then the if-else statement looks like this:

```
if (condition) {
   statement A1
   statement A2
   statement A3
} else {
   statement B1
   statement B2
   statement B3
}
```

The group of statements between a "{" and a "}" is treated as one statement by the if and else.

It is well known that vectors form the basic building block of R programming. The ifelse() function is a vectorized version of the if/else statement. It is much faster than applying the same function to each element of the vector individually. The syntax has the form ifelse(test_condition, x, y), and returns a vector of the length of its longest argument, with elements `x[i]` if `test_condition[i]` is true, otherwise `x[i]`.

A.4.2   Iterations

An **iteration** is, in principle, a loop or repeated execution of a set of statements. All modern programming languages provide special constructs that allow for the repetition of instructions. For non-array-oriented programming languages, like C, FORTRAN, even a simple matrix multiplication needs three nested loops over rows, columns, and indices. In contrast, R is array-oriented, and we can make our programming much more efficient using operations on vectors or arrays. So even though they are helpful, we would like to avoid loops whenever possible. Instead, we try to use vectorized statements or functions like `apply()`. For example, let's construct a 4×5 matrix, and we want to obtain the mean of each column. We can use the `apply()` function like the following:

```
X <- matrix(rnorm(20), nrow = 4, ncol = 5)
apply(X, 2, mean)
```

```
## [1]  0.2897 -0.3833 -0.1730 -0.2631 -0.1935
```

The "apply" family members include

- `apply`: applies over the margins of an array (e.g., the rows or columns of a matrix);
- `lapply`: applies over an object and return list;
- `sapply`: applies over an object and return a simplified object (an array) if possible;
- `vapply`: similar to sapply, but we can specify the type of object returned by the iterations.

We will treat the different forms of loops, `for` (execute a loop a fixed number of times) and `while` (execute a loop while a condition is true), in more detail in the following.

A.4.2.1   The For Loop

For loops in R are most commonly used for iterating over the elements of an object, such as list, vector, etc. The following shows the template of a **for statement** loop over all elements in a list or vector:

```
for (variable in sequence) {R commands}
```

Try these examples at the R prompt.

```
for (i in 1:3)
  print(i)

fhosp <- list("Hospital", TRUE, 95)
for (i in fhosp) print(i)
```

A.4.2.2   The While Loop

If we need a loop for which we don't know in advance how many iterations there will be, we can use the **while statement**. It works like this:

```
while (condition) {R commands}
```

While loops begin by testing a condition. We can construct a condition in the same way we did for an if statement above. If it is true, then R executes the loop body. For example, suppose we are interested in calculating the sum over 1, 2, 3 until the sum is larger than 1000. The following while loop can help us achieve the goal.

```
n <- 0                       # the iteration counter
sum.so.far <- 0              # store the added values
while (sum.so.far <= 1000){
  n <- n + 1
  sum.so.far <- sum.so.far + n
}
sum.so.far
## [1] 1035
n
## [1] 45
```

A.4.2.3   R Break and Next Statements

In R programming, we can use the break or next statement to alter a normal looping sequence.

A **break statement** is used inside a loop to exit the iterations immediately (regardless of what iteration the loop may be on), and flow the control outside of the loop. The syntax of a break statement is:

```
if (test expression) {
  break
}
```

For example,

```
for (iter in 1:100) {
  print(iter)
      if(iter > 30) {
              ## Stop loop after 30 iterations
              break
      }
}
```

A **next statement** is useful when we want to skip the current iteration of a loop without terminating it. The syntax of next statement is:

```
if (test condition) {
  next
}
```

For example,

```
for (iter in 1:100) {
        if (iter <= 10) {
                ## Skip the first 10 iterations
                next
        }
        print(iter)
}
```

B

Appendix B

B.1 COVID-19 Data and Factors Integrated from Multiple Sources

Since the first infected case reported in December 2019, the outbreak of coronavirus disease (COVID-19) has unfolded around the world. At the time of writing this book, coronavirus has infected more than 76 million people and killed over 900,000 people in the US. While essential research studies in measuring and modeling COVID-19 and its impact are underway, reliable and accurate datasets are vital for scientists to conduct related research and for decision-makers to make better decisions. Every day, several volunteer groups and organizations work really hard on collecting COVID-19 data from all the counties and states in the US. There are four primary sources, including (i) the *New York Times* (NYT)[1], (ii) the COVID Tracking Project at *The Atlantic* (Atlantic)[2], (iii) the data repository by the Center for Systems Science and Engineering (CSSE) from Johns Hopkins University (JHU)[3], and (iv) USAFacts[4].

The IDDA R package accompanying this book contains the COVID-19 epidemic data up to the county level in the US, covering about 3,200 county-equivalent areas from 50 US states and the District of Columbia. The package also includes control measures and other local information, such as demographic characteristics, socioeconomic status, healthcare infrastructure, and some other essential factors.

B.1.1 Epidemic Data

The daily counts of cases and deaths of COVID-19 are crucial for understanding how this pandemic is spreading. Thanks to the contribution of the data science communities across the world, multiple sources are providing

---

[1] https://github.com/nytimes/covid-19-data
[2] https://covidtracking.com
[3] https://github.com/CSSEGISandData/COVID-19
[4] https://usafacts.org/visualizations/coronavirus-covid-19-spread-map

the COVID-19 data with different precision and focus. In this book, we consider the reported cases from the aforementioned four sources: the *NYT*, *The Atlantic*, the JHU, and the USAFacts. To clean the data, we first fetch data from the above four sources and compile them into the same format for further comparison and cross-validation.

B.1.1.1   State Level

The state-level epidemic data contains the following variables:

- `State`: name of the state. There are 48 mainland US states and the District of Columbia.

- `XYYYY.MM.DD`: cumulative infection or death cases related to the date of `YYYY.MM.DD`. `YYYY`, `MM` and `DD` represent the year, month and day, respectively. It starts from `X2020.01.21`. For example, the variable `X2020.01.22` is either the number of infected or death cases in a certain state (`State`) on January 22, 2020.

- `Infected`: cumulative infected cases in the thousands.

- `Death`: cumulative death count.

- `Y.Infected`: daily new infected cases in the thousands.

- `Y.death`: daily new death count.

Among those variables, the variable `State` can be used as the key for data merge.

B.1.1.2   County Level

The county-level epidemic data contains the following variables:

- `ID`: county-level Federal Information Processing System (FIPS) code, which uniquely identifies the geographic area. The number has five digits, of which the first two are the FIPS code of the state to which the county belongs.

- `County`: name of the county matched with `ID`. There are about 3,200 counties and county-equivalents (e.g., independent cities, parishes, boroughs) in the US.

- `State`: name of the state matched with `ID`. There are 50 states and the District of Columbia in the US.

- `XYYYY.MM.DD`: cumulative infection or death cases related to the date of `YYYY.MM.DD`. `YYYY`, `MM` and `DD` represent the year, month and day, respectively. For example, the variable `X2020.01.22` is either the number of infected or death cases in a certain county (`County`) on January 22, 2020.

As the key of these data frames, the variable ID can be used for future data merges.

B.1.2 Other Factors

When analyzing the reported cases of COVID-19, many other factors may also contribute to the temporal or spatial patterns; see the discussions in Wang et al. (2021a). For example, local features, such as demographic and socioeconomic factors, can dramatically affect the course of the epidemic, and thus, the spread of the disease could differ substantially across different geographical regions. Therefore, these datasets are also supplemented with the population information at the county level in our repository. We further classify these factors into the following six groups.

B.1.2.1 Policy Data

In a race to stunt the spread of COVID-19, federal, state, and local governments have issued various executive orders. Government declarations are used to identify the dates that different jurisdictions implemented various social distancing policies (emergency declarations, school closures, bans on large gatherings, limits on bars, restaurants, and other public places, the deployment of severe travel restrictions, and "stay-at-home" or "shelter-in-place" orders). For example, President Trump declared a state of emergency on March 13, 2020, to enhance the federal government's response to confront COVID-19. Since late April of 2020, all 50 states in the US began to reopen successively. A state is categorized as "reopening" once its stay-at-home order lifts, or once reopening is permitted in at least one primary sector (restaurants, retail stores, personal care businesses), or once reopening is permitted in a combination of smaller sectors.

B.1.2.2 Demographic Characteristics

In the category of demographic characteristics, we consider the factors describing racial, ethnic, sexual, and age structures. These variables are extracted from the 2010 Census[5], and 2010–2018 American Community Survey (ACS) Demographic and Housing Estimates[6].

Specifically, we include the following variables:

- AA_PCT: the percent of the population who identify as African American.

---

[5]https://data.census.gov/cedsci/table?q=urban%20rate&hidePreview=false&tid=DECENNIAL
SF12010.H2&vintage=2010

[6]https://data.census.gov/cedsci/table?q=population%20density&hidePreview=false&tid=A
CSDP1Y2018.DP05&vintage=2018

- HL_PCT: the percent of the population who identify as Hispanic or Latino.

- Mortality: the five-year (1998–2002) average mortality rate, measured by the total counts of deaths per 100, 000 population in a county.

- Old_PCT: the percent of aged people (age $\geq$ 65 years).

- PD_log: the logarithm of the population density per square mile of land area.

- Pop_log: the logarithm of local population.

- Sex_ratio: the ratio of male over female.

B.1.2.3 Healthcare Infrastructure

The dataset describing the local characteristics also incorporates features related to the healthcare infrastructure, including the percent of persons under 65 years without health insurance, the local government expenditures for health per capita, and total bed counts per 1,000 population:

- EHPC: the local government expenditures for health per capita.

- NHIC_PCT: the percent of persons under 65 years without health insurance.

- TBed: the total bed counts per 1,000 population.

B.1.2.4 Socioeconomic Status

Diverse socioeconomic factors collected from 2005–2009 ACS five-year estimates[7] are also included in the county-level datasets:

- Affluence: social affluence generated by factor analysis from HighIncome, HighEducation, WCEmployment and MedHU.

- Disadvantage: concentrated disadvantage obtained by factor analysis from HHD_PAI_PCT, HHD_F_PCT and Unemployment_PCT.

- Gini: the Gini coefficient, a measure for income inequality and wealth distribution in economics.

- HHD_PAI_PCT: the percent of the households with public assistance income.

- HHD_F_PCT: the percent of households with female householders and no husband present.

- HEducation_PCT: the percent of the population aged 25 years or older with a bachelor's degree or higher.

---

[7]https://data.census.gov/cedsci/table?q=gini%20coefficient&hidePreview=false&tid=ACS
DT1Y2018.B19083&vintage=2018

- `HIncome_PCT`: the percent of families with annual incomes higher than $75,000.

- `MedHU`: the median value of owner-occupied housing units.

- `Unemployment_PCT`: civilian labor force unemployment rate.

B.1.2.5 Environmental Factor

The dataset also includes environmental factors that might affect the spread of epidemics significantly, such as the following variables:

- `PropertyCrime`: the total number of property crimes per 1,000 population.

- `ResidStability`: the percent of the population that resided in the same house for one year or more.

- `UrbanRate`: urban rate.

- `ViolentCrime`: the total number of violent crimes per 1,000 population.

B.1.2.6 Geographic Information

The `longitude` and `latitude` of the geographic center for each county in the US are available in Gazetteer Files[8]. We also refer to the GitHub repository of plotly[9] for county-level data such as `Region` and `Division`.

B.1.3 Datasets

- `CA.county.ts`: a $20,010 \times 6$ data frame with columns `ID`, `County`, `State`, `DATE`, `Death`, and `Y.death`. Each row of the data frame represents the record of one county in California on a specific date.

- `county.top10`: a 10×5 data frame with columns `ID`, `County`, `State`, `Infection`, `Death`. The rows represent the top 10 counties with the largest cumulative infected count on December 31, 2020.

- `county.top10.long`: a $3,560 \times 8$ data frame with columns `ID`, `County`, `State`, `Date`, `Count`, `type`, `Count_lb`, and `Count_ub`. This data frame focuses on the ten counties with the largest observed cumulative infected count on December 11, 2020, along with the observed cumulative infected counts from January 22, 2020 to December 11, 2020, and the predicted counts, and the 95% upper and lower bound for December 12, 2020 to January 11, 2021.

---

[8]https://www2.census.gov/geo/docs/maps-data/data/gazetteer/2019_Gazetteer
[9]https://github.com/plotly/datasets/blob/master/geojson-counties-fips.json

- `counties1@data`: a $3,109 \times 18$ data frame with columns id, GEO_ID, STATE, COUNTY, NAME, LSAD, CENSUSAREA, state_name, density, state_name_ns. Region, Division, pop, DATE, population, Infected, Death, and Infect_risk. Each row of the data frame represents a county, with the infected counts, death counts and infection risk on December 31, 2020. Note that id represents the FIPS code for the counties. NAME is the name of the counties. state_name and state_name_ns are the names of the states with and without space, respectively.

- `counties2@data`: a $3,221 \times 21$ data frame with columns id, GEO_ID, STATE, COUNTY, NAME, LSAD, CENSUSAREA, state_name, density, state_name_ns, Region, Division, pop_state, population, X2020.03.01, X2020.04.01, X2020.05.01, X2020.06.01, X2020.07.01, X2020.08.01, and X2020.09.01. Each row of the data frame shows whether a county had COVID-19 related control policies in effect on the first day of March to September, 2020. Note that id represents the FIPS code for the counties. NAME is the name of the counties. state_name and state_name_ns are the names of the states with and without space respectively.

- `D.county`: a $3,104 \times 349$ data frame with columns ID, County, State and variables from X2020.12.31 to X2020.01.21. Each row of the data frame represents the time series of the cumulative deaths for one county.

- `D.state`: a 49×347 data frame with column State and variables from X2020.12.31 to X2020.01.21. Each row of the data frame represents the time series of the cumulative deaths for one state.

- `features.county`: a 3136×36 data frame with columns ID, County, State, FIPS_C, FIPS_S, Motality, AA_PCT, HL_PCT, Gini, Affluence, HighIncome, Hincome_PCT, EduAttain, OccupAdv, MedHU, Disadcantage, PublicAssistance, HHD_PAI_PCT, HHD_F_PCT, Unemployment_PCT, ViolentCrime, PropertyCrime, ResidStability, UrbanRate, NHTC_PCT, EHPC, Latitude, Longitude, Sex_ratio, dPop_ml2, Pop_log, prop_old, BED_SUM, PD_log, Tbed, and Old_PCT. Each row of the data frame represents one county with 31 feature variables.

- `features.state`: a 51×4 data frame with columns State, Region, Division and pop. Each row of the data frame represents one state with its region, division, and population.

- `fore`: a forecast class object, which contains the two-week-ahead prediction and prediction intervals of the daily new infected cases (from January 1 to January 14, 2021) based on the historical time series of daily new infected cases in Iowa.

- `I.county`: a $3,104 \times 349$ data frame with columns ID, County, State and variables from X2020.12.31 to X2020.01.21. Each row of the data frame represents the time series of infected cases for one county.

- `I.state`: a 49×347 data frame with column `State` and variables from `X2020.12.31` to `X2020.01.21`. Each row of the data frame represents the time series of infected cases for one state.

- `pop.county`: a $3,142 \times 4$ data frame with columns `ID`, `County`, `State`, and `population`. Each row of the data frame represents one county along with its population.

- `pop.state`: a 51×2 data frame with columns `State` and `population`. Each row of the data frame represents one state along with its population.

- `state.long`: a $16,905 \times 7$ data frame with columns `State`, `Region`, `Division`, `pop`, `DATE`, `Infected` and `Death`. Each row of the data frame represents one state on a specific date.

- `state.ts`: a $16,856 \times 9$ data frame with columns `State`, `Region`, `Division`, `pop`, `DATE`, `Infected`, `Death`, `Y.Infected`, and `Y.death`. Each row of the data frame represents one state on a specific date.

- `states1@data`: a 49×11 data frame with columns `id`, `name`, `density`, `name_ns`. `Region`, `Division`, `pop`, `DATE`, `population`, `Infected`, and `Death`. Each row of the data frame represents a state, with the infected counts, death counts and infection risk on December 31, 2020. Note that `id` represents the FIPS code for the states. `state_name` and `state_name_ns` are the names of the states with and without space, respectively.

B.2 CDC FluView Portal Data

The Centers for Disease Control and Prevention (CDC) in the US maintains the FluView portal[10], where we may obtain national, regional, and state-level influenza information.

The CDC FluView portal provides in-season and past seasons' national, regional, and state-level outpatient illness and viral surveillance data from ILINet (Influenza-like Illness Surveillance Network) and WHO/NREVSS (National Respiratory and Enteric Virus Surveillance System).

The number and percent of patients suffering from an influenza-like illness (ILI), defined as fever, cough, and/or sore throat without a known cause other than influenza, would vary depending on time, season, age, and geography.

The "cdcfluview" R package pulls flu season statistics from the FluView site (a weekly influenza surveillance report) and FluView Interactive (an online

---

[10] `https://gis.cdc.gov/grasp/fluview/fluportaldashboard.html`

application that allows for more in-depth exploration of influenza surveillance data). The function `ilinet()` can get state, regional, or national influenza information from the CDC.

```
ilinet(region = c("national", "hhs", "census", "state"), years = NULL)
```

- `region`: four types of regions to retrieve data (`national`, `hhs`, `census`, and `state`);
- `years`: a vector of years, which corresponds to CDC flu season. For example, `year = 2010` is used for CDC flu season 2010–2011. Default is `NULL` to retrieve all years since 1997.

For example, we can load CDC FluView Portal Data for the national level since 1997 as follows:

```
library(cdcfluview)
ili.usa <- ilinet(region = "national", years = NULL)
# str(ili.usa)
```

The output of the `ilinet()` function consists of

- `age_X_XX`: the number of outpatient with ILI by age group (`age_0_4` for 0–4 years, `age_25_49` for 25 – 49 years, `age_25_64` for 25 – 64 years, `age_5_24` for 5 – 24 years, `age_50_64` for 50 – 64 years, and `age_65` for $\geq$ 65 years);
- `ilitotal`: the number of outpatient visits for ILI each week;
- `num_of_providers`: the number of outpatient healthcare providers around the country report data to the CDC;
- `region`: National for `national`, 10 HHS Regions (Region 1 to Region 10) for `hhs`, 9 Census Regions (East North Central, East South Central, Mid-Atlantic, Mountain, New England, Pacific, South Atlantic, West North Central, and West South Central) for `census`, and 55 states and territories (50 states and District of Columbia, New York City, Virgin Islands, Puerto Rico, and the Northern Mariana Islands) for `state`;
- `year`: the year of data;
- `region_type`: National, HHS Regions, Census Regions, States;
- `total_patients`: the total number of outpatients visits each week;
- `unweighted_ili`: the percent of outpatient visits for ILI each week;
- `week`: the week of data;

- `week_start`: the start date of a week;

- `weighted_ili`: the percent of outpatient visits for ILI, weighted by state population, each week.

C

Appendix C

As in the majority of scientific disciplines, dates and times play an important role in infectious disease learning. This appendix provides an short introduction to the main R packages that deal with date and date/time data.

C.1 Classes: R Dates and Times

In R, dates are represented by the "Date" class, and date/time values are represented by the "POSIXct" or the "POSIXlt" class.

- "Date" class: stores time as the number of days since UNIX epoch on 1970-01-01, with negative values for earlier dates.

- "POSIXct" class: stores date/time values as the number of seconds since January 1, 1970.

- "POSIXlt" class: stores date/time values as a list with elements for the second, minute, hour, day, month, and year, among others.

The "POSIXct" and "POSIXlt" classes are ISO-compliant data objects supporting date/times with time zones and assorted calendar adjustments. These classes are especially useful when time zone manipulation is important.

R provides several options for dealing with date and time data. In the following, we will introduce how to use functions from both the "base" R package and the "lubridate" package to work with date-time data classes.

C.2 Formatting Date and Date/Time Variables

The general rule for date/time data in R is to use the simplest technique possible. For a dataset that only contains dates (without times), `as.Date()` will usually be the best choice as it handles dates directly.

To coerce a variable to date, we can use the `as.Date()` function. We can use the `unclass()` function to see the internal integer representation of the date in R.

```
report_date <- as.Date("2020-01-21")
report_date
## [1] "2020-01-21"
class(report_date)
## [1] "Date"
# Number of days after January 1, 1970.
unclass(report_date)
## [1] 18282
```

The `as.Date()` function allows one to provide a format argument. The default is `%Y-%m-%d`, which is year-month-day.

```
as.Date("21-01-2020", format - "%d-%m-%Y")
## [1] "2020-01-21"
as.Date("X2020.01.21", "X%Y.%m.%d")
## [1] "2020-01-21"
```

The following symbols can be used with the `format()` function for dates.

Symbol	Description	Example
%d	Day of the month (01-31)	05, 24
%a	Abbreviated weekday	Mon, Sun
%A	Unabbreviated weekday	Monday, Sunday
%m	Month (00-12)	02, 11
%b	Abbreviated month	Jan, Jul
%B	Unabbreviated month	January, July
%j	Day of the year (001-366)	152, 258
%y	2-digit year without century	08, 20
%Y	4-digit year with century	2020
%w	Weekday (0-6) [Sunday is 0]	1, 6
%W	Week (00-53)	08, 20

Let's consider the dataset `I.state` in the IDDA package as an example. Recall that the column names "XYYYY.MM.DD" are not names of variables but values of a variable, which contains the values of the cumulative infected count.

In the following, we first pivot the column names into a new variable DATE, and then we convert the DATE values to the "Date" class.

```
# devtools::install_github('FIRST-Data-Lab/IDDA')
library(IDDA)
data(I.state)
library(dplyr)
library(tidyr)

I.state.long <- I.state %>%
  # Select the columns to pivot
  pivot_longer(cols = X2020.12.31:X2020.01.21,
               names_to = "DATE",
               values_to = "Infected") %>%
  # Convert the DATE cell values to Date format
  mutate(DATE = as.Date(DATE, "X%Y.%m.%d"))

head(I.state.long, 2)
## # A tibble: 2 x 3
##    State     DATE         Infected
##    <chr>     <date>          <int>
## 1 Alabama 2020-12-31       361226
## 2 Alabama 2020-12-30       356820
```

If we have an object containing both date and time, we can use as.POSIXct() or as.POSIXlt() to store both date and time. As illustrated in the following, the as.POSIX* functions help convert an object to one of the two classes.

```
report_time <- as.POSIXct("01/21/2020 15-24-30",
                          format = "%m/%d/%Y %H-%M-%S")
report_time
## [1] "2020-01-21 15:24:30 EST"
class(report_time)
## [1] "POSIXct" "POSIXt"
```

C.3   Creating Data/Time Objects in R

R has many useful built-in dates that we can use. For example, the base R functions Sys.time() and Sys.Date() return the system's idea of the current date with and without time. The "lubridate" package is also able to return these values using the functions lubridate::today() for the current date and lubridate::now() for the current time.

We can generate a sequence of "Date" object using the seq() function. For example, to generate a vector of seven dates, starting on January 19, 2020, with an interval of one day between them, we can use:

```
report_week <- seq(as.Date("2020-01-19"), by = "days", length = 7)
report_week
## [1] "2020-01-19" "2020-01-20" "2020-01-21" "2020-01-22"
## [5] "2020-01-23" "2020-01-24" "2020-01-25"
```

Using the "lubridate" package is very similar. The only difference is "lubridate" changes the way we specify the arguments related to the "Date" class in the seq() function. Note that "lubridate" is part of the "tidyverse" package. We have seen other "tidyverse" packages in Chapter 2. If the "tidyverse" package has been installed, then we should have the "lubridate" package installed already.

```
#install.packages("lubridate")
library(lubridate)
report_week <- seq(ymd("2020-01-19"), by = "days", length = 7)
report_week
## [1] "2020-01-19" "2020-01-20" "2020-01-21" "2020-01-22"
## [5] "2020-01-23" "2020-01-24" "2020-01-25"
```

C.4   Parsing Date and Time

While creating date-time objects, we can specify different formats using the conversion specification; however, we often need to deal with data imported from a database. We need to parse the date/time from the data imported in such cases. This section will focus on parsing date/time from character data. Both the base R package and the "lubridate" package offer functions to parse date and time, and we will explore a few of them in this section.

C.4.1 Date-time Conversion to and from Character Using Base R Functions

The `strptime()` function in the base R package allows us to create a date and/or time object from a given string in a certain format. The following is the template for basic R syntax:

```
# Convert character to time object
strptime(character, format)
# Convert time object to character
strftime(time, format)
```

For example, as demonstrated below, the `strftime(report_week)` function converts the Date vector object `report_week` to a character vector object.

```
class(report_week)
## [1] "Date"
str_report_week <- strftime(report_week, format = "%Y-%m-%d")
class(str_report_week)
## [1] "character"
```

Section C.2 illustrated how to pivot a wide table with column names being the date to a wide table. The following example demonstrates the inverse transformation of a long data table with values of a Date object to a wide table with character strings of the date being the variables. Before the transformation, we first convert the date values into character strings. We can choose a format that we like to style a date, for example, "X.YYYY.MM.DD" as shown in the code below.

```
I.state.wide <- I.state.long %>%
    # Replace the Date column DATE with its formatted character strings
    mutate(DATE = strftime(DATE, format = "X.%Y.%m.%d")) %>%
    pivot_wider(names_from = DATE, values_from = Infected)
head(I.state.wide, 2)
```

C.4.2 Parsing Date and Time Using "lubridate"

To parse strings into a "Date" object with the "lubridate" package, we simply provide the order in which year, month, and day appear in the dates, then arrange "y," "m," and "d" in the same order, which gives the name of the lubridate function to parse the date as follows.

```
library(lubridate)
ymd("2020-01-21")
## [1] "2020-01-21"
mdy("January 21st, 2020")
## [1] "2020-01-21"
dmy("21-Jan-2020")
## [1] "2020-01-21"
```

Another approach is to use the lubridate::parse_date_time() function, which parses an input vector into the "POSIXct" date-time object. It differs from base::strptime() in two respects. First, it allows specification of the order in which the formats occur without including separators and the % prefix. Second, it allows the user to specify several format-orders to handle heterogeneous date-time character representations.

```
# Heterogeneous date-times
report_date <- c("20-01-21", "20200121", "20-01 21")
parse_date_time(report_date, "ymd")
## [1] "2020-01-21 UTC" "2020-01-21 UTC" "2020-01-21 UTC"
parse_date_time(report_date, "y m d")
## [1] "2020-01-21 UTC" "2020-01-21 UTC" "2020-01-21 UTC"
parse_date_time(report_date, "%y%m%d")
## [1] "2020-01-21 UTC" "2020-01-21 UTC" "2020-01-21 UTC"
```

Note that the lubridate::parse_date_time() function always assigns the "Co-ordinated Universal Time" (UTC) time zone.

C.5 Setting and Extracting Information

The "lubridate" package offers a number of generic functions that work on dates and times to help us extract pieces of dates and/or times, for example, `second`, `minute`, `hour`, `day`, `wday`, `week`, `month`, `year`, and `tz`.

```r
report_date <- as.Date("2020-01-21")
month(report_date)   # month number
## [1] 1
day(report_date)      # day (number) of the month
## [1] 21
wday(report_date)    # day number of the week (1-7)
## [1] 3
```

We can apply these functions to a Date column. For example, we can extract the month, day and year information and create new columns `Month`, `Day` and `Year` for the `I.state.long` data.

```r
I.state.MDY <- I.state.long %>%
  mutate(Month = month(DATE), Day = day(DATE), Year = year(DATE))
```

```r
head(I.state.MDY, 2)
## # A tibble: 2 x 6
##    State    DATE        Infected Month   Day  Year
##    <chr>    <date>        <int> <dbl> <int> <dbl>
## 1 Alabama 2020-12-31     361226    12    31  2020
## 2 Alabama 2020-12-30     356820    12    30  2020
```

We can also extract time components from a datetime object or column, for example, we can use the following code to extract hour, minute and second of `report_datetime`.

```r
report_datetime <- ymd_hm("2020-03-01 14:45")
hour(report_datetime)     # extract hour
```

```r
## [1] 14
```

```
minute(report_datetime)    # extract minute
```

```
## [1] 45
```

```
second(report_datetime)    # extract second
```

```
## [1] 0
```

C.5.1   Epidemiological Calendar

The epidemiological calendar is used on a daily basis in disease surveillance. To carry out epidemiological surveillance activities, we sometimes need to group disease outbreaks or epidemiological events around a given period of time. An **epidemiological week** (epi week or a CDC week) is simply a standardized method of counting weeks to compare data year after year in epidemiological surveillance.

Epi weeks start on a Sunday and end on a Saturday. The first epi week of the year ends on the first Saturday of January, provided that it falls at least four or more days into the month, even if it means that this first week starts in December.

The "lubridate" R package supports for the following week or year numbering systems:

- week(): returns the number of complete seven-day periods that have occurred between the date and January 1st, plus one;

- isoweek(): returns the week as it would appear in the ISO 8601 system, which uses a reoccurring leap week;

- epiweek(): returns the US CDC version of the epidemiological week;

- isoyear(): returns years according to the ISO 8601 week calendar;

- epiyear(): returns years according to the epidemiological week calendars.

```
library(lubridate)
I.state.MDYW <- I.state.MDY %>%
  mutate(Week = week(DATE),
```

```
        EpiWeek = epiweek(DATE), IsoWeek = isoweek(DATE),
        EpiYear = epiyear(DATE), IsoYear = isoyear(DATE)) %>%
  select(-DATE)

head(I.state.MDYW, 2)
## # A tibble: 2 x 10
##   State   Infected Month  Day  Year  Week EpiWeek
##   <chr>      <int> <dbl> <int> <dbl> <dbl>   <dbl>
## 1 Alabama   361226    12    31  2020    53      53
## 2 Alabama   356820    12    30  2020    53      53
## # ... with 3 more variables: IsoWeek <dbl>,
## #   EpiYear <dbl>, IsoYear <dbl>
```

C.6   Merging Separate Date Information

The as.Date() function can be used in conjunction with the paste() function
to combine columns separate date information (e.g., year, month, day) into
one date variable.

```
I.state.long <- I.state.MDYW %>%
  mutate(date = as.Date(paste(Year, Month, Day, sep = "-"))) %>%
  select(State, date, Infected)
head(I.state.long, 2)
## # A tibble: 2 x 3
##   State   date       Infected
##   <chr>   <date>        <int>
## 1 Alabama 2020-12-31   361226
## 2 Alabama 2020-12-30   356820
```

C.7   Date Calculations in R

Both "Date" and "POSIXct" R objects are represented by simple numerical
values, which makes calculation with time and date objects very straightforward. R uses the underlying numerical values to make calculations, and then
converts the result back to human-readable time information again.

We can increment and decrement "Date" objects, or do actual calculations with them:

```
report_date <- as.Date("2020-01-21")
report_date + 14
## [1] "2020-02-04"
as.Date("2020-03-12") - as.Date("2020-02-27")
## Time difference of 14 days
range(I.state.long$date)
## [1] "2020-01-21" "2020-12-31"
```

The difftime(date1, date2, units) function in R can be used to calculate the time difference between date1 and date2 in the required units such as days, weeks, months, years, etc.

```
date1 <- as.Date("2020-03-16")
date2 <- as.Date("2020-01-21")
difftime(date1, date2, units = "days")
## Time difference of 55 days
difftime(date1, date2, units = "weeks")
## Time difference of 7.857 weeks
```

We can sort the Date object easily in R. For example, if we would like to sort the data I.state.long from the earliest to the most recent date, we can apply the dplyr::arrange() function like the following:

```
# Earliest (top) to most recent (bottom) order
I.state.long <- I.state.long %>%
  arrange(date)
head(I.state.long)
## # A tibble: 6 x 3
##    State        date         Infected
##    <chr>        <date>          <int>
## 1 Alabama      2020-01-21          0
## 2 Arizona      2020-01-21          0
## 3 Arkansas     2020-01-21          0
## 4 California   2020-01-21          0
## 5 Colorado     2020-01-21          0
## 6 Connecticut  2020-01-21          0
```

Bibliography

Abbott, S., Hellewell, J., Thompson, R. N., Sherratt, K., Gibbs, H. P., Bosse, N. I., Munday, J. D., Meakin, S., Doughty, E. L., Chun, J. Y., et al. (2020). Estimating the time-varying reproduction number of SARS-CoV-2 using national and subnational case counts. *Wellcome Open Research*, 5(112):112.

Altieri, N., Barter, R. L., Duncan, J., Dwivedi, R., Kumbier, K., Li, X., Netzorg, R., Park, B., Singh, C., Tan, Y. S., Tang, T., Wang, Y., Zhang, C., and Yu, B. (2021). Curating a COVID-19 data repository and forecasting county-level death counts in the united states. *Harvard Data Science Review*.

Arik, S. O., Li, C.-L., Yoon, J., Sinha, R., Epshteyn, A., Le, L. T., Menon, V., Singh, S., Zhang, L., Yoder, N., et al. (2020). Interpretable sequence learning for COVID-19 forecasting. *arXiv preprint arXiv:2008.00646*.

Arora, P., Kumar, H., and Panigrahi, B. K. (2020). Prediction and analysis of COVID-19 positive cases using deep learning models: A descriptive case study of India. *Chaos, Solitons & Fractals*, 139.

Assimakopoulos, V. and Nikolopoulos, K. (2000). The theta model: A decomposition approach to forecasting. *International Journal of Forecasting*, 16(4):521–530.

Barreto, M. L., Teixeira, M. G., and Carmo, E. H. (2006). Infectious diseases epidemiology. *Journal of Epidemiology & Community Health*, 60(3):192–195.

Bates, J. M. and Granger, C. W. (1969). The combination of forecasts. *Journal of the Operational Research Society*, 20(4):451–468.

Bettencourt, L. M. and Ribeiro, R. M. (2008). Real time Bayesian estimation of the epidemic potential of emerging infectious diseases. *PloS One*, 3(5):e2185.

Brauer, F. (2008). Compartmental models in epidemiology. In *Mathematical epidemiology*, pages 19–79. Springer.

Brauer, F., van den Driessche, P., and Wu, J. (2008). *Mathematical epidemiology*, volume 1945. Springer.

Breiman, L. (2001). Statistical modeling: The two cultures. *Statist. Sci.*, 16(3):199–231. With comments and a rejoinder by the author.

Brockwell, P. J. and Davis, R. A. (2016). *Introduction to time series and forecasting.* Springer Texts in Statistics. Springer, [Cham], third edition.

Brown, R. G. (1959). *Statistical forecasting for inventory control.* New York: McGraw-Hill.

Castro1, L., Fairchild, G., Michaud, I., and Osthus, D. (2020). COFFEE: COVID-19 forecasts using fast evaluations and estimation. https://covid-19.bsvgateway.org/static/COFFEE-methodology.pdf.

Chen, L.-P., Zhang, Q., Yi, G. Y., and He, W. (2021). Model-based forecasting for Canadian COVID-19 data. *PloS One*, 16(1):e0244536.

Cleveland, R. B., Cleveland, W. S., McRae, J. E., and Terpenning, I. (1990). Stl: A seasonal-trend decomposition. *Journal of Official Statistics*, 6(1):3–73.

Cleveland, W. S. (1979). Robust locally weighted regression and smoothing scatterplots. *Journal of the American Statistical Association*, 74(368):829–836.

Cori, A., Ferguson, N. M., Fraser, C., and Cauchemez, S. (2013). A new framework and software to estimate time-varying reproduction numbers during epidemics. *American Journal of Epidemiology*, 178(9):1505–1512.

Daley, D. J. and Gani, J. (2001). *Epidemic modelling: An introduction.* Cambridge Studies in Mathematical Biology. Cambridge University Press.

Dancho, M. and Vaughan, D. (2020). *anomalize: Tidy anomaly detection.* R package version 0.2.2.

De Boor, C. (1978). *A practical guide to splines.* Spriger-Verlag.

De Livera, A. M., Hyndman, R. J., and Snyder, R. D. (2011). Forecasting time series with complex seasonal patterns using exponential smoothing. *Journal of the American Statistical Association*, 106(496):1513–1527.

Dicker, R. and Gathany, N. (1992). *Principles of epidemiology.* Public Health Service, Centre for Disease Control and Prevention (CDC), Atlanta, second edition.

Dietz, K. and Heesterbeek, J. (2002). Daniel Bernoulli's epidemiological model revisited. *Mathematical Biosciences*, 180(1-2):1–21.

Durojaye, M. and Ajie, I. (2017). Mathematical model of the spread and control of Ebola virus disease. *Appl. Math*, 7:23–31.

En'Ko, P. (1989). On the course of epidemics of some infectious diseases. *International Journal of Epidemiology*, 18(4):749–755.

Ferguson, N. M. et al. (2020). Impact of non-pharmaceutical interventions (NPIs) to reduce COVID-19 mortality and healthcare demand. Report 9, Imperial College London, London, United Kingdom. https://doi.org/10.25561/77482.

Ferland, R., Latour, A., and Oraichi, D. (2006). Integer-valued GARCH process. *Journal of Time Series Analysis*, 27(6):923–942.

Gardner, E. S. and Mckenzie, E. (1985). Forecasting trends in time series. *Management Science*, 31(10):1237–1246.

Hadley, W. (2016). *Ggplot2: Elegant graphics for data analysis*. Springer.

Held, L., Hens, N., O'Neill, P. D., and Wallinga, J. (2020). *Handbook of infectious disease data analysis*. New York: Chapman and Hall/CRC.

Hoertel, N., Blachier, M., Blanco, C., Olfson, M., Massetti, M., Rico, M. S., Limosin, F., and Leleu, H. (2020). A stochastic agent-based model of the SARS-CoV-2 epidemic in France. *Nature Medicine*, 26(9):1417–1421.

Holt, C. C. (1957). Forecasting seasonals and trends by exponentially weighted moving averages. *Pittsburgh, Penn: Carnegie Institute of technology, Graduate School of Industrial Administration*.

Hurvich, C. M. and Tsai, C.-L. (1989). Regression and time series model selection in small samples. *Biometrika*, 76(2):297–307.

Hyndman, R. J. and Athanasopoulos, G. (2018). *Forecasting: Principles and practice*. OTexts.

Jewell, N. P. (2003). *Statistics for epidemiology*. CRC Press.

Keeling, M. J. and Rohani, P. (2008). *Modeling infectious diseases in humans and animals*. Princeton University Press.

Kermack, W. O. and McKendrick, A. G. (1927). A contribution to the mathematical theory of epidemics. *Proceedings of the Royal Society of London. Series A, Containing Papers of a Mathematical and Physical Character*, 115(772):700–721.

Kim, M., Gu, Z., Yu, S., Wang, G., and Wang, L. (2021). Methods, challenges, and practical issues of COVID-19 projection: A data science perspective. *Journal of Data Science*, 19(2).

Lawson, A. B., Banerjee, S., Haining, R. P., and Ugarte, M. D. (2016). *Handbook of spatial epidemiology*. CRC Press.

Lessler, J. and Cummings, D. A. (2016). Mechanistic models of infectious disease and their impact on public health. *American Journal of Epidemiology*, 183(5):415–422.

Miranda, G. H., Baetens, J. M., Bossuyt, N., Bruno, O. M., and De Baets, B. (2019). Real-time prediction of influenza outbreaks in Belgium. *Epidemics*, 28:100341.

Mojeeb, A., Adu, I. K., and Yang, C. (2017). A simple SEIR mathematical model of malaria transmission. *Asian Research Journal of Mathematics*, pages 1–22.

Pfeiffer, D., Robinson, T. P., Stevenson, M., Stevens, K. B., Rogers, D. J., Clements, A. C., et al. (2008). *Spatial analysis in epidemiology*, volume 142. Oxford University Press.

Porta, M. (2014). *A dictionary of epidemiology*. Oxford University Press.

Rahmandad, H. and Sterman, J. (2008). Heterogeneity and network structure in the dynamics of diffusion: Comparing agent-based and differential equation models. *Management Science*, 54(5):998–1014.

Rai, B., Shukla, A., and Dwivedi, L. K. (2021). Estimates of serial interval for COVID-19: A systematic review and meta-analysis. *Clinical Epidemiology and Global Health*, 9:157–161.

Roda, W. C., Varughese, M. B., Han, D., and Li, M. Y. (2020). Why is it difficult to accurately predict the COVID-19 epidemic? *Infectious Disease Modelling*, 5:271–281.

Ruppert, D., Wand, M. P., and Carroll, R. J. (2003). *Semiparametric regression*. Cambridge University Press.

Sanche, S., Lin, Y. T., Xu, C., Romero-Severson, E., Hengartner, N., and Ke, R. (2020). High contagiousness and rapid spread of severe acute respiratory syndrome coronavirus 2. *Emerging Infectious Diseases*, 26(7):1470.

Sievert, C. (2020). *Interactive web-based data visualization with R, plotly, and shiny*. CRC Press.

Srivastava, N., Hinton, G., Krizhevsky, A., Sutskever, I., and Salakhutdinov, R. (2014). Dropout: a simple way to prevent neural networks from overfitting. *The Journal of Machine Learning Research*, 15(1):1929–1958.

Sujath, R., Chatterjee, J. M., and Hassanien, A. E. (2020). A machine learning forecasting model for COVID-19 pandemic in India. *Stochastic Environmental Research and Risk Assessment*, 34:959–972.

Tang, Z. and Fishwick, P. A. (1993). Feedforward neural nets as models for time series forecasting. *ORSA Journal on Computing*, 5(4):374–385.

Tashman, L. J. (2000). Out-of-sample tests of forecasting accuracy: an analysis and review. *International Journal of Forecasting*, 16(4):437–450.

Venables, W. N., Smith, D. M., Team, R. D. C., et al. (2009). *An introduction to R*. Citeseer.

Wallinga, J. and Teunis, P. (2004). Different epidemic curves for severe acute respiratory syndrome reveal similar impacts of control measures. *American Journal of Epidemiology*, 160(6):509–516.

Wang, G., Gu, Z., Li, X., Yu, S., Kim, M., Wang, Y., Gao, L., and Wang, L. (2021a). Comparing and integrating US COVID-19 data from multiple sources with anomaly detection and repairing. Available at https://doi.or g/10.1080/02664763.2021.1928016.

Wang, L., Wang, G., Li, X., Yu, S., Kim, M., Wang, Y., Gu, Z., and Gao, L. (2021b). Modeling and forecasting COVID-19. *Notices of the American Mathematical Society*, 68(4):585–595.

Wickham, H. and Grolemund, G. (2016). *R for data science: Import, tidy, transform, visualize, and model data*. O'Reilly Media, Inc.

Winters, P. R. (1960). Forecasting sales by exponentially weighted moving averages. *Management Science*, 6(3):324–342.

Wood, S. N., Pya, N., and Säfken, B. (2016). Smoothing parameter and model selection for general smooth models. *Journal of the American Statistical Association*, 111(516):1548–1563.

Zou, D., Wang, L., Xu, P., Chen, J., Zhang, W., and Gu, Q. (2020). Epidemic model guided machine learning for COVID-19 forecasts in the United States. *medRxiv*.

Index

For Product Safety Concerns and Information please contact our
EU representative GPSR@taylorandfrancis.com Taylor & Francis
Verlag GmbH, Kaufingerstraße 24, 80331 München, Germany